AF410855

Droplet Microfluidics

Droplet Microfluidics

Editors

Eric Brouzes
Siran Li

MDPI • Basel • Beijing • Wuhan • Barcelona • Belgrade • Manchester • Tokyo • Cluj • Tianjin

Editors
Eric Brouzes
Stony Brook University
USA

Siran Li
Cold Spring Harbor
USA

Editorial Office
MDPI
St. Alban-Anlage 66
4052 Basel, Switzerland

This is a reprint of articles from the Special Issue published online in the open access journal *Micromachines* (ISSN 2072-666X) (available at: https://www.mdpi.com/journal/micromachines/special_issues/Droplet_Based_Microfluidics).

For citation purposes, cite each article independently as indicated on the article page online and as indicated below:

LastName, A.A.; LastName, B.B.; LastName, C.C. Article Title. *Journal Name* **Year**, *Volume Number*, Page Range.

ISBN 978-3-0365-0184-0 (Hbk)
ISBN 978-3-0365-0185-7 (PDF)

Contents

About the Editors

Eric Brouzes, Ph.D., Associate Professor, Department of Biomedical Engineering at Stony Brook University. Dr. Brouzes graduated from the ESPCI-Paris Tech in Physics before obtaining his Ph.D. from Universite Paris 7 for his work performed at the Institut Curie on the early development of Drosophila Melanogaster. He completed a post-doc in the laboratory of Prof. Perrimon at Harvard Medical School and started a collaboration with Raindance Technologies where he discovered droplet microfluidics. He later joined the company as a senior scientist before consulting for the start-up Metagenomix. He moved back to academia as a Research Assistant Professor in the Department of Biomedical Engineering at Stony Brook University where he was recently promoted to Associate Professor with tenure.

Siran Li , Ph.D., Research Assistant Professor, Cold Spring Harbor Laboratory. Dr. Li earned his B.S. from Peking University, and Ph.D. in Electrical Engineering at Princeton University. He then joined Cold Spring Harbor Laboratory as a postdoctoral fellow under Dr. Michael Wigler before being promoted to Research Assitant Professor. His research focuses on novel single-cell sequencing technology development.

Editorial

Editorial for the Special Issue on Droplet Microfluidics

Eric Brouzes [1,2,3,4,*] and Siran Li [5,*]

1   Department of Biomedical Engineering, Stony Brook University, Stony Brook, NY 11794, USA
2   Laufer Center for Physical and Quantitative Biology, Stony Brook University, Stony Brook, NY 11794, USA
3   Cancer Center, Stony Brook School of Medicine, Stony Brook, NY 11794, USA
4   Institute for Engineering Driven Medicine, Stony Brook University, Stony Brook, NY 11794, USA
5   Cold Spring Harbor Laboratory, Cold Spring Harbor, NY 11724, USA
*   Correspondence: eric.brouzes@stonybrook.edu (E.B.); siranli@cshl.edu (S.L.)

Received: 1 December 2020; Accepted: 2 December 2020; Published: 8 December 2020

Emulsions, which are collections of immiscible droplets, have elicited scientific and commercial interests for decades. However, droplet microfluidics has only recently emerged due to advances in microfluidics and physical chemistry. Droplet microfluidics uniquely allows encapsulation into monodisperse and stable droplets. Those properties represent the two pillars of the technology because they enable manipulating droplets as independent micro-reactors both on-chip and off-chip. Droplets are generated and processed with high precision in microfluidic circuits that can include different modules in series.

Since its inception, droplet microfluidics has inspired numerous research topics, ranging from fluid hydrodynamics to biology, chemistry, and material sciences. The field is indeed highly interdisciplinary as it combines fluid dynamics, microfabrication, physical-chemistry, chemistry, and biology. Droplet microfluidics has also garnered commercial success through applications such as digital PCR or single-cell genomics.

This special issue is a collection of eight articles that perfectly illustrate the breadth of this vibrant field. It represents a testimony to the creativity involved in the development of droplet microfluidics. Six articles cover automation and templating capabilities of droplet microfluidics for biological, chemical, and material sciences applications; the last two articles present novel modules that proposes a solution to the limited mixing in single phase microfluidics and that provides new sorting abilities for biological applications.

Droplet microfluidics promises workflow automation with improved throughput and reduced reagent consumption. The paper by Lindong Weng et al. [1] reviews the advantages of droplet microfluidics to screen large libraries for directed evolution. Directed evolution is arguably the initial impetus behind the development of droplet microfluidics because droplets readily link the genotype (coding sequence) to the phenotype (protein) via encapsulation. The paper also provides a thorough review of the microfluidic modules used to manipulate and process droplets at high throughput and thus serves as a perfect introduction to the technology. Mark Davies et al. [2] present a droplet-based commercial system's theory and practice to generate combinatorial drug libraries for high throughput screening. The combinatorial power of droplets combined with reagent consumption reduction surpasses conventional technologies such as robotics for large and complex screening strategies. Hoon Suk Rho et al. [3] introduce a system that combines droplet microfluidics with on-chip valves, two technologies that have been recently combined together. Their low throughput platform enables fine droplet control and the design of complex chemical and biological reactions. Finally, Nan Shi et al. [4] demonstrate a screening platform with automated calibration that provides real-time correction for fluorescence drift. The repeated calibration is essential to improve the quality of analytical assays, especially those that rely on expensive reagents.

Microfluidics droplets can be used as templates for manufacturing original microscale objects. For instance, Jianhua Guo et al. [5] harness osmotic pressure to precisely tune the thickness of the ultrathin shell of microparticles manufactured at high throughput. Employing unusual fluids, Qingming Hu et al. [6] demonstrate the generation of liquid metal droplets numerically and experimentally. These two papers exemplify the ability to use droplet microfluidics to create new materials with novel properties and applications in very diverse fields.

The approach of Xiaoyu Jia et al. [7] contrasts with the typical use of droplets as microreactors by utilizing them as actuators. They overcome the limited mixing of single-phase microfluidics by creating chaotic convective turbulent flow with air bubbles. Finally, Chandler Dobso et al. [8] use a photo-reactive surfactant to tag droplets of interest that can then be passively sorted out due to the induced change in interfacial energy. Unlike conventional sorting methods, where the sorting decision is made just before sorting, this functionality temporaly uncouples observation, decision algorithm, and sorting. This uncoupling opens up the ability of long term observation of droplets before sorting.

We are grateful to all the authors who submitted their papers to this special issue dedicated to droplet microfluidics. We are also indebted to all the reviewers who dedicated their time and helped improve the quality of the submitted papers.

Conflicts of Interest: The author declares no conflict of interest.

References

1. Weng, L.; Spoonamore, J.E. Droplet Microfluidics-Enabled High-Throughput Screening for Protein Engineering. *Micromachines* **2019**, *10*, 734. [CrossRef] [PubMed]
2. Davies, M.; Abubaker, M.; Bible, L. A Flexible, Microfluidic, Dispensing System for Screening Drug Combinations. *Micromachines* **2020**, *11*, 943. [CrossRef] [PubMed]
3. Rho, H.S.; Gardeniers, H. Microfluidic Droplet-Storage Array. *Micromachines* **2020**, *11*, 608. [CrossRef] [PubMed]
4. Shi, N.; Easley, C.J. Programmable μChopper Device with On-Chip Droplet Mergers for Continuous Assay Calibration. *Micromachines* **2020**, *11*, 620. [CrossRef] [PubMed]
5. Guo, J.; Hou, L.; Hou, J.; Yu, J.; Hu, Q. Generation of Ultra-Thin-Shell Microcapsules Using Osmolarity-Controlled Swelling Method. *Micromachines* **2020**, *11*, 444. [CrossRef] [PubMed]
6. Hu, Q.; Jiang, T.; Jiang, H. Numerical Simulation and Experimental Validation of Liquid Metal Droplet Formation in a Co-Flowing Capillary Microfluidic Device. *Micromachines* **2020**, *11*, 169. [CrossRef] [PubMed]
7. Jia, X.; Che, B.; Jing, G.; Zhang, C. Air-Bubble Induced Mixing: A Fluidic Mixer Chip. *Micromachines* **2020**, *11*, 195. [CrossRef] [PubMed]
8. Dobson, C.; Zielke, C.; Pan, C.W.; Feit, C.; Abbyad, P. Method for Passive Droplet Sorting after Photo-Tagging. *Micromachines* **2020**, *11*, 964. [CrossRef] [PubMed]

Publisher's Note: MDPI stays neutral with regard to jurisdictional claims in published maps and institutional affiliations.

 *micromachines*

Article

Method for Passive Droplet Sorting after Photo-Tagging

Chandler Dobson †, Claudia Zielke †, Ching W. Pan, Cameron Feit and Paul Abbyad *

Department of Chemistry and Biochemistry, Santa Clara University, Santa Clara, CA 95053, USA;
cdobson@scu.edu (C.D.); czielke@scu.edu (C.Z.); chingwei_pan@urmc.rochester.edu (C.W.P.);
cfeit@scu.edu (C.F.)
* Correspondence: pabbyad@scu.edu
† Both authors contributed equally to this work.

Received: 21 September 2020; Accepted: 26 October 2020; Published: 28 October 2020

Abstract: We present a method to photo-tag individual microfluidic droplets for latter selection by passive sorting. The use of a specific surfactant leads to the interfacial tension to be very sensitive to droplet pH. The photoexcitation of droplets containing a photoacid, pyranine, leads to a decrease in droplet pH. The concurrent increase in droplet interfacial tension enables the passive selection of irradiated droplets. The technique is used to select individual droplets within a droplet array as illuminated droplets remain in the wells while other droplets are eluted by the flow of the external oil. This method was used to select droplets in an array containing cells at a specific stage of apoptosis. The technique is also adaptable to continuous-flow sorting. By passing confined droplets over a microfabricated trench positioned diagonally in relation to the direction of flow, photo-tagged droplets were directed toward a different chip exit based on their lateral movement. The technique can be performed on a conventional fluorescence microscope and uncouples the observation and selection of droplets, thus enabling the selection on a large variety of signals, or based on qualitative user-defined features.

Keywords: microfluidics; droplet microfluidics; sorting; passive sorting; photo-tag; droplet array

---

1. Introduction

Droplet microfluidics confines chemicals or cellular reagents in picoliter droplets transported in inert oil and is well suited for large-scale chemical or biological analysis [1,2]. The selection of droplets is a basic operation for applications such as cell screening, [3,4] enzyme directed evolution [5], and metabolic engineering [6,7]. In active sorting, [8] droplets that exceed a fluorescence signal threshold, are synchronized with an electric field, [9,10] magnetic field, [11] acoustic wave, [12] pneumatic pressure [13], or light [14] to direct selected droplets to a predetermined channel exit. For example, fluorescence-activated droplet sorting (FADS) [10] uses electric fields for robust selection at rates as high as 30 kHz [15].

Although powerful, these techniques are endpoint measurements that require the observation and active sorting to be performed close to each other in both time and space. Droplets are observed in series and the order of droplets must be maintained to ensure that the active sorting component selects the correct droplets. This limits the sorting criteria in two major ways. The fluorescence signals must generally be measured just prior to active sorting. Therefore, droplets cannot be selected based on the time evolution of signals or signals that would develop or saturate much earlier in time. Second, in-line sorting also limits the observation time of droplets to approximately the droplet sorting time. For fast sorting, the selection signal is based on moving droplets integrated for milliseconds or less. The signal must be easily interpretable without user intervention. Selection is thus primarily based on fluorescence signals, [10,16] and, in rare occasions, fluorescence lifetime [17] and absorbance [18].

It excludes weak signals that would require longer integration or signals that would require specialized equipment or user interpretation. Examples of techniques that would be difficult to integrate with active droplet sorting due to a combination of instrumentation and long signal integration include FT-IR [19] and Raman detection [20], UV-Vis absorbance, fluorescence anisotropy, [21], and droplet radiofluidics [22]. Cell selection based on subtle differences in cell morphology, important criteria for determining cell type and state, often requires user interpretation. Kinetic measurements in droplets, such as fast kinetics [23] and Michaelis-Menton kinetic measurements [24], are also difficult to integrate with current active droplet sorting technology.

Droplet arrays [25–27] provide a platform for extended observation of droplets with selection either by mechanical means with pneumatic valves [28] or via photoactivation, producing an air droplet that selectively pushes the droplet out of the well [29,30]. Both methods have potential drawbacks. Valves increase the complexity and control elements of the device. Photoactivation requires a UV or NIR laser and can cause localized heating that may damage droplet content. Moreover, as selection is based on droplet location or address in the array, both the droplet observation and selection must be performed on the array. This can limit the type of operations or steps that can be performed on the droplets.

We present here a simple and inexpensive method that can be performed on a conventional fluorescence microscope that uncouples the observation and selection of droplets allowing the selection on a larger variety of signals.

We have recently developed a passive technology, Sorting by Interfacial Tension (SIFT), [31,32] that sorts droplets based on differences in interfacial tension that is linked to droplet pH. The technique was used previously to sort droplets containing enzymes [31] and cells [32,33]. In both these cases, the change in the pH was due to processes incurred by droplet content not user control. In the case of enzymes, the change of pH was from the production of an acidic byproduct. For cells, the release of protons associated with glycolysis induced a pH change. Hence, these applications did not allow the selection of an arbitrary droplet by the user.

Photoacids have the property of becoming more acidic upon light excitation. Droplets containing the photoacid, 8-hydroxypyrene-1,3,6-trisulfonic acid trisodium salt (HPTS), henceforth called by its common name pyranine, were excited, leading to a decrease in pH and a concurrent increase in droplet interfacial tension. This provides a method to "tag" droplets of interest for latter passive sorting based on differences in interfacial tension.

The "tagging" of droplets for selection uncouples the observation step from the sorting step. This is a departure from most sorting techniques, such as FADS, where the observation and sorting are required to occur close together in both time and space. The observation step can be as long as needed. This opens the door for selection of droplets from signals that require specialized equipment, long integration time, or user intervention (i.e.,: absorbance, kinetic data, cell morphology, IR, Raman, fluorescence lifetime, anisotropy, and droplet radiofluidics). The technique is easy to perform and requires only a conventional fluorescence microscope. It is flexible as the tagging can be performed anywhere on the chip. This simplifies device design and the integration of measurement and sorting within the same chip. The technique does not require the detector, active sorting element, and related electronics of FADS. Thus, it can greatly decrease the cost and complexity of an analysis.

The mechanism and application of the sorting method are presented here. First, the change in pH of droplets as a function of the duration of light excitation is characterized. The technique is used to select individual droplets within a droplet array. Spatial separation is demonstrated by exciting or tagging specific droplets long before the downstream passive sorting of these same droplets. The technique is also used to select desired cells in a droplet array based on user-determined apoptosis stage. By uncoupling the measurement and sorting of droplets, this novel method presents a simple way to increase the variety and complexity of signals that can be used as the basis for droplet selection.

2. Materials and Methods

2.1. Microfluidic Device

Polydimethylsiloxane (PDMS) microfluidic chips with channel depth modulations were fabricated using the dry-film photoresist soft lithography technique described by Stephan et al. [34]. The technique enabled rapid prototyping of multi-level structures. The PDMS chips were then plasma-bonded to a glass slide. To render the internal channel surface hydrophobic, Novec™ 1720 Electronic Grade Coating (3M) was flowed into the microchannel and the chip was heated for 30 min at 150 °C. The surface treatment prevented wetting and contact of the aqueous droplets with the channel walls.

2.2. Beads

A solution of 2 μm fluorescent red beads (Sigma-Aldrich, Milwaukee, WI, USA) in a 1:20,000 ratio was prepared in 2.5 mM PBS buffer supplemented with Optiprep (15% *v/v*, Fresenius Kabi Norge AS for Axis-Shield PoCAS, Oslo, Norway), Pluronic F-68 (1% *w/w*, Affymetrix Inc., Maumee, OH, USA), and 2 mM pyranine (AAT Bioquest Inc., Sunnyvale, CA, USA). Optiprep was added to limit bead and cell sedimentation in the tubing and droplets, whereas Pluronic F-68 increased cell viability and droplet stability. Pyranine is a photoacid and used in these experiments to selectively mark then sort droplets. In addition, it serves as fluorescent ratiometric probe for pH analysis of droplets, based on a calibration curve of droplets of known pH.

2.3. Cells

Jurkat, Clone E6-1 (ATCC TIB-152, human acute T-cell leukemia) cells were grown in ATCC-formulated RPMI-1640 Medium, containing 10% fetal bovine serum (HyClone, GE Healthcare Life Sciences, Logan, UT, USA) and 1% penicillin-streptomycin (Gibco, Life Technologies Corporation, Grand Island, NY, USA). The growing conditions were 37 °C and 5% CO_2.

A population of Jurkat cells was harvested through centrifugation, washed with PBS, and resuspended in PBS. The harvested cells were kept for 1–3 days on the bench to induce apoptosis. On the day of experiment, cells were centrifuged, resuspended in 2.5 mM PBS buffer supplemented with Optiprep, Pluronic F-68 and pyranine as described above. Both, pH and osmolality (determined with Vapro Vapor Pressure Osmometer 5520, Wescor, ELITech Biomedical Systems, Logan, UT, USA) of the solutions were analyzed and adjusted to match physiological values (pH 7.4; 280–320 mOsmol). Around 2 drops/mL of Invitrogen Propidium Iodide ReadyProbes™ Reagent (Fisher Scientific) were added to visualize apoptosis based on membrane integrity under fluorescent light.

On the day of experiment, a second population of Jurkat cells was harvested and prepared for on-chip experiments following the same procedure as described above. Both apoptotic and freshly harvested cells were mixed in a 1:1 ratio before injection onto the chip to ensure a variety of cell viability stages for the experiment. To reduce the amount of double occupied droplets, the cell density was analyzed with a Cellometer Auto T4 Bright Field Cell Counter (Nexcelcom Bioscience LLC, Lawrence, MA, USA) and adjusted to about 1×10^6 cells/mL prior to injection.

2.4. Measurements

The temperature of the chip was maintained at 37 °C using a microscope heating stage with control module with temperature feedback (CHS-1 heating plate with TC-324C temperature controller, Warner Instruments). Fluid flow was controlled using computer-controlled syringe pumps (Nemesys, Cetoni). Droplets were produced in 0.1% Picosurf in perfluorinated oil, Novec 7500 (Sphere Fluidics, Cambridge, UK). Another perfluorinated oil with dissolved surfactant, Droplet Generation Oil for Probes (Bio-Rad, Hercules, CA, USA), hereby called QX100, was used to push out droplets from the array and within the sorting rail region.

Images and videos were taken on an inverted fluorescence microscope (Olympus IX-51) with a shuttered LED fluorescence excitation source (Spectra-X light engine, Lumencor, Beaverton, OR,

USA) and a high-speed camera (VEO-410, Vision Research, Wayne, NJ, USA). The excitation source had individual addressable LEDs that included violet (395 nm BP 25 nm), blue (436 nm BP 28 nm), and green (561 nm BP 14 nm) light. Based on manufacturer specifications, each LED had approximately 300 mW of power. The excitation source was coupled to an Arduino (Arduino LLC, Scarmagno, Italy) to rapidly alternate between different colored LEDs or to use two simultaneously using simple TTL triggering. For photo-tagging droplets, a 40× objective was used with a partially closed iris to limit the area of excitation. The microscope filter cube contained a dual-edge dichroic mirror (Di03-R488/561-t1-25 × 36, Semrock, IDEX Health & Science LLC Rochester, NY, USA) and dual-band emission filter (FF01-523/610-25, Semrock) that enabled transmission of both pyranine and the red fluorescence of the beads or propidium iodide.

Droplets containing fluorescent beads were identified using green excitation light. Cell apoptosis stage was identified using a 40× magnification. Live cells (spherical) and necrotic cells (decomposed) were identified by morphology using bright field. Late apoptosis cells were distinguished from early apoptosis cells by propidium iodide fluorescence using green excitation. Images and videos were analyzed using ImageJ [35].

3. Results and Discussion

In droplet microfluidics, reagents are encapsulated in aqueous droplets transported in inert oil enabling increases in throughput in applications such as digital PCR, directed evolution, drug screening, and single-cell analysis [4,5,10,16]. A critical step for many applications is the selection and recovery of droplets of interest. Sorting is performed by first observing the droplet, then using an active component to direct desired droplets to a different chip exit.

We have recently developed a label-free technique to sort droplets based on pH. The technique is based on the observation that the interfacial tension of droplets can be very sensitive to droplet pH for a specific oil and surfactant combination. In particular, for droplets in the external oil and surfactant combination of QX100, interfacial tension increases from 12 to 24 dyn/cm as pH decreases from 8 to 6 [32]. These pendant-drop measurements were performed with QX100 solution diluted 100-fold with perfluorinated oil. Droplets of undiluted QX100 quickly detached from the needle. The interfacial tension of droplets in undiluted QX100, as used in experiments, is much lower and estimated to be a few dyn/cm from calculations [31]. The source of the pH sensitivity of QX100 is suspected to be due to a change in protonation state of the surfactant; however, the proprietary nature of the product renders it difficult to confirm the exact mechanism [31].

In a technique dubbed Sorting by Interfacial Tension (SIFT), differences in interfacial tension of droplets provides a handle to sort droplets based on pH. The method utilizes flattened droplets, confined by the top and bottom of the channel. Droplets encounter a microfabricated trench, or rail, of increased channel height (rail geometry in Supplemental Figure S1). The droplets expand into the rail, decreasing their surface area and energy. To leave the rail, droplets need to be squeezed again. Droplets of low interfacial tension enter the rail but are pushed off due to the drag of the external oil flow. The oil flow is insufficient to push droplets of higher interfacial tension off the rail. These droplets follow the tapered rail oriented at 45 degrees relative to the direction of flow. These droplets leave at a different lateral position as compared to droplets of low interfacial tension. Using different channel exits, droplets of different interfacial tension and hence pH are sorted. The technique has been utilized for label-free sorting of different enzymes based on activity [31] and of cells based on glycolysis [32,33]. In these applications, the pH change was induced by droplet content rather than external control.

The SIFT technique is expanded here by using light to induce a selective pH change in a droplet containing a fluorescent dye. The fluorophore pyranine is a water-soluble dye that is used as a staining agent, pH indicator, and model system for excited-state proton transfer (ESPT) [36,37]. It is cell membrane impermeant and considered biocompatible [38]. When excited, pyranine gives off a characteristic yellow-green fluorescence. It is a ratiometric pH indicator as the excitation spectrum is modulated by pH, shifting from blue (440 nm) to violet (405 nm) with decreasing pH [39].

It was observed that the excitation of a solution of pyranine leads to a decrease in solution pH as measured with a pH meter. No change in pH was observed in the absence of pyranine. The change in pH was characterized as a function of irradiation time with light for a linear array of stationary droplets (Figure 1). The droplets were irradiated simultaneously with both violet (395 nm BP 25 nm) and blue (436 nm BP 28 nm) light using a 40× objective. After irradiation, the fluorescence was monitored with either blue or violet light with a 10× objective. For the blue excitation channel (Figure 1A), the fluorescence intensity decreases as a function of irradiation time. In contrast, for violet excitation (Figure 1B), the fluorescence intensity remains mostly constant with irradiation time.

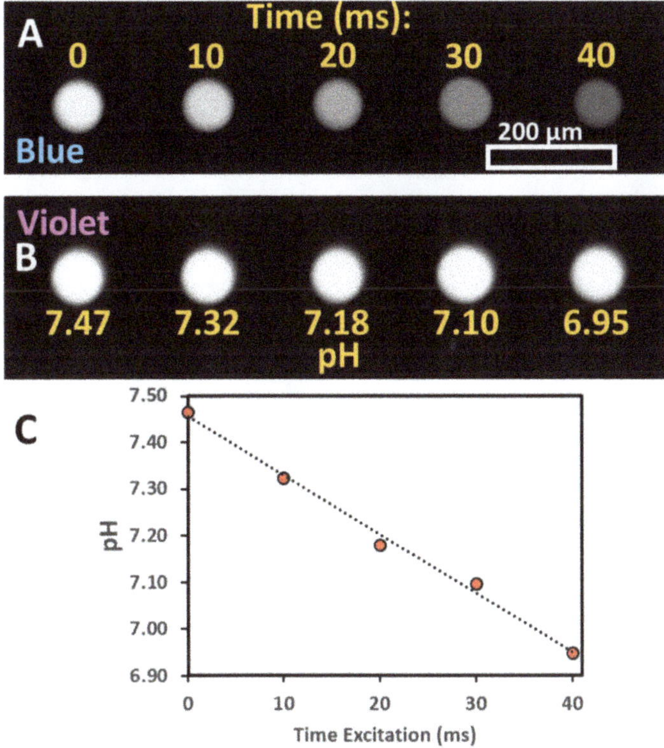

Figure 1. Linear array of droplets with increasing irradiation times from left to right. (**A**) Fluorescence from blue excitation (436 nm). (**B**) Fluorescence from violet excitation (395 nm). pH of droplets as determined from blue to violet fluorescence intensity ratio based on calibration curve as shown in Supplemental Figure S2. (**C**) Droplet pH with increasing irradiation time with fit to a linear regression.

The pH of droplets after irradiation can be determined by the ratio of blue to violet excitation of pyranine using a calibration curve (Supplemental Figure S2). The decrease in pH is correlated with irradiation time (Figure 1C) and is observed to be mostly linear. For long irradiation times (>100 ms), the ratio of violet to blue was no longer proportional to excitation time. It was found that the change of pH was faster with the irradiation of droplets with a combination of violet and blue LED lights than either alone (data not shown). The monitoring of droplet pH could itself induce a pH change. To limit this effect, the exposure time was limited and a lower numerical aperture objective was used for monitoring (10× magnification with a numerical aperture of 0.30) than for irradiation (40× magnification with a numerical aperture of 0.60).

It is worth noting that the ratio of violet to blue fluorescence for a given pH was the same regardless of whether a change in pH was induced by light or the addition of a strong acid (Supplemental Figure S3). This shows that pyranine can serve both to induce a pH change as well as to report droplet pH.

As a photoacid, pyranine shows a marked increase in acidity in the excited state with a pKa that decreases from 7.4 to 0.4 in the excited state [40]. Thus, the photoexcitation of pyranine would lead to a release of a proton. This proton would be expected to return upon relaxation to the ground state. However, intense excitation can lead to photooxidation and radical reactions of the pyranine in the excited state, [41,42] and lead to the observed change in droplet pH. The destruction of fluorophore is consistent with the constant fluorescence intensity upon violet excitation (Figure 1B) as an increase would otherwise be expected at lower pH for a constant pyranine concentration.

The decrease in droplet pH upon irradiation can be used to select specific droplets of a droplet array as shown in Figure 2 and Video S1. Figure 2A shows the channel geometry as well as the inlets and outlets of the microfluidic device. The droplets at an initial pH of 7.46 were anchored in an array of wells using a technique called Rails and Anchors, [26] where confined droplets expand into wells (Figure 2B). Droplets were approximately the size of the wells. To exit the well, droplets need to be squeezed from their height in the well (50 μm) to the height of the channel (25 μm). In the presence of QX100, interfacial tension increased with decreasing droplet pH. Therefore, droplets of higher pH are pushed out of the wells at lower external flow rates (Supplemental Figure S4). The flow required for droplet ejection also provided an estimate of droplet pH. This was useful for longer irradiations and lower pH when ratiometric readings of pyranine fluorescence did not provide an accurate estimate of droplet pH.

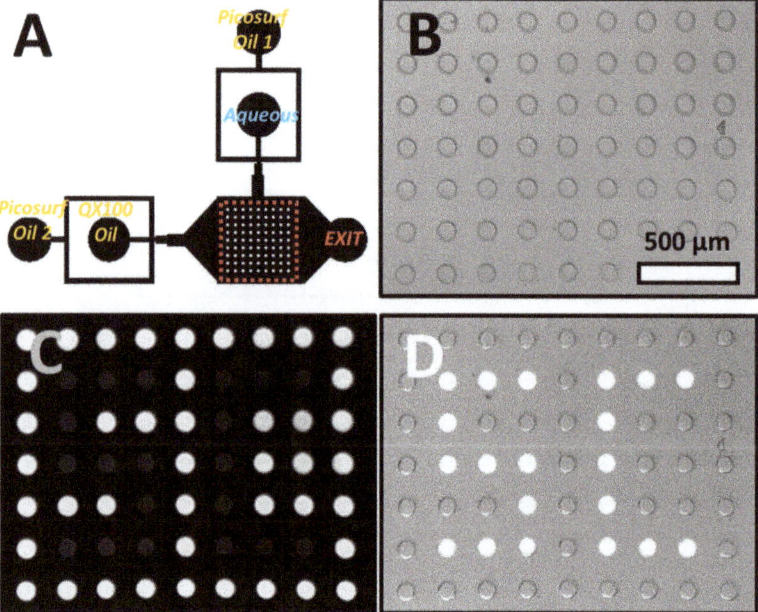

Figure 2. (**A**) Array device channel geometry. The location of the droplet array is highlighted by a dashed red square. (**B**) Bright-field image of droplet array. All wells are occupied with droplets that are approximately the same size as the wells. (**C**) Fluorescence from blue excitation. Irradiated droplets have lower fluorescence intensity. (**D**) Droplet array after the elution of non-irradiated droplets. Fluorescence excitation light was used in combination with bright-field image to increase droplet visibility.

Selected droplets were irradiated for about a second as described above with violet and blue light using a 40× objective. This excitation led to a decrease of over 1 pH unit as estimated from the external flow required for elution. The droplets were noticeably dimmer when excited with blue light (Figure 2C). Next, QX100 was flowed at high flow rates, 80–100 µL/min, into the device. The droplets at higher pH were quickly displaced from the wells. However, for the irradiated droplets at lower pH and hence higher interfacial tension, the hydrodynamic drag was insufficient to push out the droplets. These droplets remained in the wells as seen in the fluorescence image in Figure 2D (letters "S" and "C" produced with droplets). Increasing the flow to a much higher rate, typically around 150 µL/min, caused the elution of all droplets in the array. The photoselection of droplets was dependent on both the presence of pyranine in the droplet and the use of QX100 oil. The absence of either led to no correlation between droplet excitation and droplet elution (Supplemental Figure S5).

By the general method described above, the observation and selection can be uncoupled in time. Droplets can be observed and studied in the array for an arbitrary length of time. The droplets of interest can then be selected, either by irradiation to remain in the array (positive selection) or through elution (negative selection) by irradiating undesired droplets.

A similar technique can also be used to separate the observation and selection in space, each performed in different regions of a microfluidic device. In contrast to the use of a droplet array, this variant, using a microfabricated rail (Figure 3A), would be more amenable to continuous sorting rather than batch processing. As a demonstration, droplets were produced of which a small percentage, <2%, contained fluorescent beads. Flow was slowed and droplets containing beads were selectively irradiated for 1 s as shown in Figure 3B, inset. This illumination induced a decrease in pH of about 1 unit from the initial pH of 7.4 based on a similar exposure in the droplet array device. As visible from the image, the illumination area was slightly larger than the droplet leading to the slight irradiation of the neighboring droplets. However, this off-target illumination was insufficient to lead to a sizable change in pH or to influence sorting.

Approximately 90–120 s after irradiation, illuminated droplets were flowed to the sorting region. The sorting region was located over 2 cm downstream from the location of droplets irradiation. As the droplets enter the sorting region, QX100 is introduced into the channel. The external oil flow and hence the drag can be controlled in the sorting region. This control is provided by an additional QX100 inlet, denoted as the Oil Entrainment Inlet in Figure 3A. Supplemental Table S1 provides typical flow rates in the device. Figure 3B and Video S2 show the sorting of droplets containing beads. Empty droplets, that were not irradiated, enter the sorting rail but are quickly pushed off by the entrainment flow and exit the chip via the unselected exit. The irradiated droplets containing beads have higher interfacial tension in the presence of QX100. The external flow of oil is insufficient to push the droplets off the rail. The droplets followed the tapered rail upwards leaving near the end point. These droplets (circled in green) exit the chip via the selected exit.

The method is able to isolate photo-tagged droplets within a large excess of droplets. In this case, a large change in pH was induced by light to ensure clearly differentiated droplet populations. However, the entrainment flow provides an independent user-defined parameter that can be finely tuned to select between droplets with smaller pH differences. For example, droplets irradiated for 50 ms and with a pH difference of 0.4 units were separated using the device as shown in Supplemental Figure S6. When droplets come in contact with QX100, they become more acidic as revealed by the dimmer droplets with blue excitation near the rail. This is likely because the surfactant in QX100 is acidic, however the proprietary nature of the product makes this difficult to confirm.

As a demonstration of an application of the technique, cells contained in droplets were selected on the basis of their stage in apoptosis through visual characterization, without fixing the cells. Cell apoptosis is programmed cell death, which is a vital part of biological cell turnover and an important process to ensure proper development. Improper apoptosis is linked to many disease states [43]. Apoptosis can be identified through distinct morphological changes of a cell. Using the technique presented, different cell states (viable, early apoptosis, late apoptosis, necrosis) can be observed, evaluated, selected, and collected for further analysis.

Using the device shown in Figure 2A, cells at different stages of cell death were encapsulated in droplets and anchored within an array. The cells were then evaluated and categorized (viable, early apoptosis, late apoptosis, necrosis) by user observation as described in the Materials and Methods (Figure 4A). A droplet containing a cell identified as early apoptosis (yellow diamond) was illuminated with light leading to a decrease in pH from 7.4 to 6.5. This change in pH can be observed as a decrease in fluorescence signal when excited with blue light (Figure 4B). QX100 was introduced into the chip at a flow rate of 120 µL/min ejecting empty droplets and droplets containing cells of different apoptosis stages from the array. The sole remaining droplet in the array contains the photo-tagged droplet containing the early apoptosis cell. This selected droplet can be ejected by increasing the flow rate even higher to 140 µL/min.

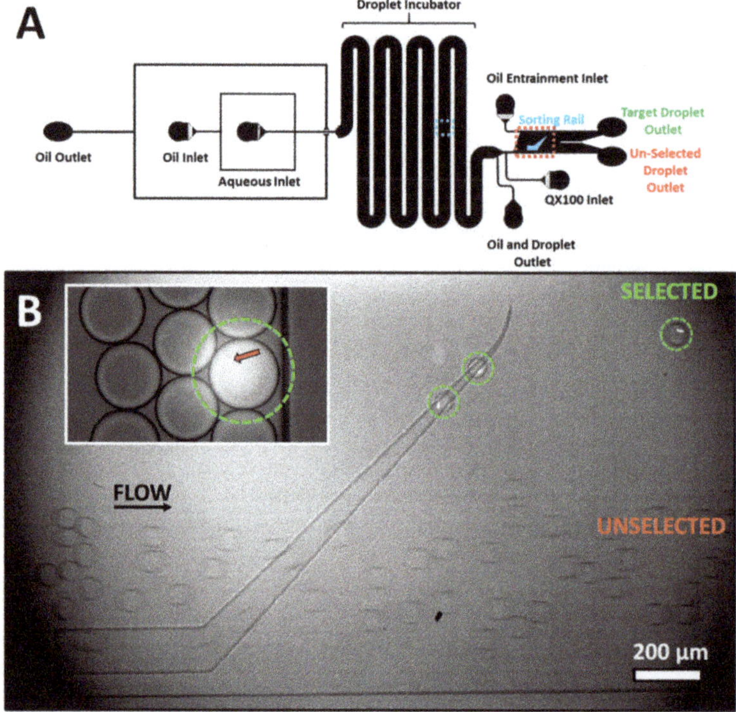

Figure 3. (**A**) Rail device channel geometry. The location of the images in (**B**) are shown. Location of droplet irradiation is indicated by a blue dashed rectangle and the sorting rail is highlighted by a dashed red rectangle. (**B**) Droplets containing fluorescent beads, circled in green, were previously irradiated with light. Irradiated droplets have lower pH and hence higher interfacial tension. They follow the rail upwards and leave at a higher lateral position toward the selected exit. Other droplets are immediately pushed off the rail by the flow of oil toward the unselected exit. Inset: A droplet circled in green was irradiated prior to sorting. Neighboring droplets are also partially irradiated. The red arrow indicates a fluorescent bead.

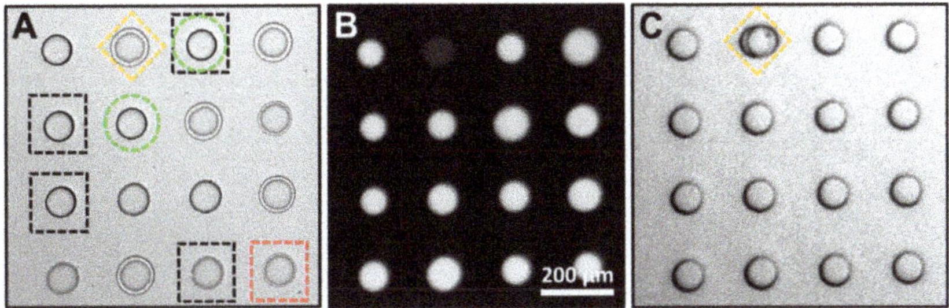

Figure 4. (**A**) Bright-field image of droplet array. Droplets containing cells were classified as viable (green circle), early apoptosis (yellow diamond), late apoptosis (red square), and necrosis (black square). (**B**) Fluorescence image with blue excitation. The droplet with an early apoptosis cell was irradiated and displays lower fluorescence intensity. (**C**) Droplet array after the elution of non-irradiated droplets. The only droplet remaining in the array contains the early apoptosis cell.

The method presents a way to select droplets and hence cells of interest. Here, the droplet with the cell of interest was irradiated and remained in the array. Alternatively, the undesired droplets can be irradiated with the desired droplets ejected. Depending on the application, this may be preferable as the excitation and decrease of the pH could damage cells. This would also limit exposure to the surfactant, QX100, that may have limited biocompatibility [32,44]. For the collection of multiple cell types, several illumination times could be used to induce different pH changes. This would allow for sequential elution of droplets by increasing flow rates.

4. Conclusions

We present a method to photoselect and then recover individual droplets by passive sorting. It can be used to select droplets in a droplet array or to direct photo-tagged droplets toward a different chip exit using a microfabricated rail. The technique is simple and inexpensive to adapt as it requires only a conventional fluorescence microscope without the need for the detector, active sorter, and related electronics used for fluorescence-activated droplet sorting (FADS). The technique uncouples the observation and sorting of droplets, thus facilitating the selection based on signals that are difficult to adapt to FADS including weak signals, kinetic data, long integration times, or qualitative features.

The method presented here can be adapted for efficiency and sensitivity. With the LED excitation source used, excitation times required here for sorting were as little as 50 ms. This time can likely be decreased with a more intense laser source. Likewise, both lowering the buffering capacity or increasing the pyranine concentration of solution will also decrease the excitation times. Pyranine, with an excitation spectrum that spans up to 500 nm, may interfere with some assays. Other photoacids, [45] could possibly be used instead to provoke a light-induced pH change. Ideal droplet tagging would cause a minimum change or perturbation to the droplet. In this initial demonstration, the change in pH was large which allowed for robust selection. However, by carefully tuning the flow conditions, consistent selection can be performed based on small pH differences, as little as 0.05 units [33]. Excitation position was adjusted here by hand using the microscope stage. A computer-controlled excitation source, using galvanometric mirrors, [46] could allow more precise and automated droplet tagging.

Supplementary Materials: Supplementary materials can be found at http://www.mdpi.com/2072-666X/11/11/964/s1.

Author Contributions: Conceptualization, C.W.P. and P.A.; investigation, C.D., C.Z., C.W.P. and C.F.; supervision, C.Z. and P.A.; writing—original draft, C.Z. and P.A; writing—review & editing, C.Z. and P.A. All authors have read and agreed to the published version of the manuscript.

Funding: This research was funded by a National Science Foundation Career Award, Grant Number 1751861, the National Institutes of Health under grant 1R15GM129674-01 and the Henry Dreyfus Teacher-Scholar Awards Program.

Acknowledgments: We acknowledge helpful discussions with Brian McNelis and Ian Carter-O'Connell of the Department of Chemistry and Biochemistry at Santa Clara University. C.D. acknowledges support from the Jean Dreyfus Lectureship for Undergraduate Institutions. C.Z. acknowledges support from the Inclusive Excellence Postdoctoral Program at Santa Clara University. We would also like to thank International Electronic Components Inc. for their generous donation of dry photoresist.

Conflicts of Interest: The authors declare no conflict of interest.

References

1. Song, H.; Chen, D.L.; Ismagilov, R.F. Reactions in droplets in microfluidic channels. *Angew. Chem. Int. Ed.* **2006**, *45*, 7336–7356. [CrossRef]
2. Guo, M.T.; Rotem, A.; Heyman, J.A.; Weitz, D.A. Droplet microfluidics for high-throughput biological assays. *Lab Chip* **2012**, *12*, 2146–2155. [CrossRef] [PubMed]
3. el Debs, B.; Utharala, R.; Balyasnikova, I.; Griffiths, A.D.; Merten, C.A. Functional single-cell hybridoma screening using droplet-based microfluidics. *Proc. Natl. Acad. Sci. USA* **2012**, *109*, 11570–11575. [CrossRef] [PubMed]
4. Brouzes, E.; Medkova, M.; Savenelli, N.; Marran, D.; Twardowski, M.; Hutchison, J.B.; Rothberg, J.M.; Link, D.R.; Perrimon, N.; Samuels, M.L. Droplet microfluidic technology for single-cell high-throughput screening. *Proc. Natl. Acad. Sci. USA* **2009**, *106*, 14195–14200. [CrossRef]
5. Agresti, J.J.; Antipov, E.; Abate, A.R.; Ahn, K.; Rowat, A.C.; Baret, J.-C.; Marquez, M.; Klibanov, A.M.; Griffiths, A.D.; Weitz, D.A. Ultrahigh-throughput screening in drop-based microfluidics for directed evolution. *Proc. Natl. Acad. Sci. USA* **2010**, *107*, 4004–4009. [CrossRef]
6. Hammar, P.; Angermayr, S.A.; Sjostrom, S.L.; van der Meer, J.; Hellingwerf, K.J.; Hudson, E.P.; Joensson, H.N. Single-cell screening of photosynthetic growth and lactate production by cyanobacteria. *Biotechnol. Biofuels* **2015**, *8*, 193. [CrossRef] [PubMed]
7. Abalde-Cela, S.; Gould, A.; Liu, X.; Kazamia, E.; Smith, A.G.; Abell, C. High-throughput detection of ethanol-producing cyanobacteria in a microdroplet platform. *J. R. Soc. Interface* **2015**, *12*, 1–7. [CrossRef]
8. Xi, H.; Zheng, H.; Guo, W.; Ganan-Calvo, A.M.; Ai, Y.; Tsao, C.-W.; Zhou, J.; Li, W.; Huang, Y.; Nguyen, N.-T.; et al. Active droplet sorting in microfluidics: A review. *Lab Chip* **2017**. [CrossRef]
9. Link, D.R.; Grasland-Mongrain, E.; Duri, A.; Sarrazin, F.; Cheng, Z.; Cristobal, G.; Marquez, M.; Weitz, D.A. Electric control of droplets in microfluidic devices. *Angew. Chem. Int. Ed.* **2006**, *45*, 2556–2560. [CrossRef]
10. Baret, J.-C.; Miller, O.J.; Taly, V.; Ryckelynck, M.; El-Harrak, A.; Frenz, L.; Rick, C.; Samuels, M.L.; Hutchison, J.B.; Agresti, J.J.; et al. Fluorescence-activated droplet sorting (FADS): Efficient microfluidic cell sorting based on enzymatic activity. *Lab Chip* **2009**, *9*, 1850–1858. [CrossRef]
11. Surenjav, E.; Priest, C.; Herminghaus, S.; Seemann, R. Manipulation of gel emulsions by variable microchannel geometry. *Lab Chip* **2009**, *9*, 325–330. [CrossRef] [PubMed]
12. Franke, T.; Abate, A.R.; Weitz, D.A.; Wixforth, A. Surface acoustic wave (SAW) directed droplet flow in microfluidics for PDMS devices. *Lab Chip* **2009**, *9*, 2625–2627. [CrossRef] [PubMed]
13. Wu, L.; Chen, P.; Dong, Y.; Feng, X.; Liu, B.F. Encapsulation of single cells on a microfluidic device integrating droplet generation with fluorescence-activated droplet sorting. *Biomed. Microdevices* **2013**, *15*, 553–560. [CrossRef] [PubMed]
14. Baroud, C.N.; de Saint Vincent, M.R.; Delville, J.-P. An optical toolbox for total control of droplet microfluidics. *Lab Chip* **2007**, *7*, 1029–1033. [CrossRef]
15. Sciambi, A.; Abate, A.R. Accurate microfluidic sorting of droplets at 30 kHz. *Lab Chip* **2015**, *15*, 47–51. [CrossRef]
16. Mazutis, L.; Gilbert, J.; Ung, W.L.; Weitz, D.A.; Griffiths, A.D.; Heyman, J.A. Single-cell analysis and sorting using droplet-based microfluidics. *Nat. Protoc.* **2013**, *8*, 870–891. [CrossRef]
17. Hasan, S.; Geissler, D.; Wink, K.; Hagen, A.; Heiland, J.J.; Belder, D. Fluorescence lifetime-activated droplet sorting in microfluidic chip systems. *Lab Chip* **2019**, *19*, 403–409. [CrossRef]
18. Gielen, F.; Hours, R.; Emond, S.; Fischlechner, M.; Schell, U.; Hollfelder, F. Ultrahigh-throughput-directed enzyme evolution by absorbance-activated droplet sorting (AADS). *Proc. Natl. Acad. Sci. USA* **2016**, *113*, E7383–E7389. [CrossRef]
19. Chan, K.L.A.; Niu, X.; Demello, A.J.; Kazarian, S.G. Generation of chemical movies: FT-IR spectroscopic imaging of segmented flows. *Anal. Chem.* **2011**, *83*, 3606–3609. [CrossRef]

20. Cecchini, M.P.; Hong, J.; Lim, C.; Choo, J.; Albrecht, T.; DeMello, A.J.; Edel, J.B. Ultrafast surface enhanced resonance raman scattering detection in droplet-based microfluidic systems. *Anal. Chem.* **2011**, *83*, 3076–3081. [CrossRef]

21. Choi, J.W.; Kang, D.K.; Park, H.; Demello, A.J.; Chang, S.I. High-throughput analysis of protein-protein interactions in picoliter-volume droplets using fluorescence polarization. *Anal. Chem.* **2012**, *84*, 3849–3854. [CrossRef] [PubMed]

22. Türkcan, S.; Nguyen, J.; Vilalta, M.; Shen, B.; Chin, F.T.; Pratx, G.; Abbyad, P. Single-Cell Analysis of [18F] Fluorodeoxyglucose Uptake by Droplet Radiofluidics. *Anal. Chem.* **2015**, 6667–6673. [CrossRef]

23. Huebner, A.M.; Abell, C.; Huck, W.T.S.; Baroud, C.N.; Hollfelder, F. Monitoring a Reaction at Submillisecond Resolution in Picoliter Volumes. *Anal. Chem.* **2011**, *83*, 1462–1468. [CrossRef]

24. Song, H.; Ismagilov, R.F. Millisecond kinetics on a microfluidic chip using nanoliters of reagents. *J. Am. Chem. Soc.* **2003**, *125*, 14613–14619. [CrossRef] [PubMed]

25. Schmitz, C.H.J.; Rowat, A.C.; Köster, S.; Weitz, D.A. Dropspots: A picoliter array in a microfluidic device. *Lab Chip* **2009**, *9*, 44–49. [CrossRef]

26. Abbyad, P.; Dangla, R.; Alexandrou, A.; Baroud, C.N. Rails and anchors: Guiding and trapping droplet microreactors in two dimensions. *Lab Chip* **2011**, *11*, 813–821. [CrossRef] [PubMed]

27. Labanieh, L.; Nguyen, T.N.; Zhao, W.; Kang, D.K. Floating droplet array: An ultrahigh-throughput device for droplet trapping, real-time analysis and recovery. *Micromachines* **2015**, *6*, 1469–1482. [CrossRef] [PubMed]

28. Jeong, H.H.; Lee, B.; Jin, S.H.; Jeong, S.G.; Lee, C.S. A highly addressable static droplet array enabling digital control of a single droplet at pico-volume resolution. *Lab Chip* **2016**, *16*, 1698–1707. [CrossRef]

29. Segaliny, A.I.; Li, G.; Kong, L.; Ren, C.; Chen, X.; Wang, J.K.; Baltimore, D.; Wu, G.; Zhao, W. Functional TCR T cell screening using single-cell droplet microfluidics. *Lab Chip* **2018**, *18*, 3733–3749. [CrossRef]

30. Han, S.H.; Choi, Y.; Kim, J.; Lee, D. Photoactivated Selective Release of Droplets from Microwell Arrays. *ACS Appl. Mater. Interfaces* **2020**, *12*, 3936–3944. [CrossRef]

31. Horvath, D.G.; Braza, S.; Moore, T.; Pan, C.W.; Zhu, L.; Pak, O.S.; Abbyad, P. Sorting by interfacial tension (SIFT): Label-free enzyme sorting using droplet microfluidics. *Anal. Chim. Acta* **2019**, *1089*, 108–114. [CrossRef] [PubMed]

32. Pan, C.W.; Horvath, D.G.; Braza, S.; Moore, T.; Lynch, A.; Feit, C.; Abbyad, P. Sorting by interfacial tension (SIFT): Label-free selection of live cells based on single-cell metabolism. *Lab Chip* **2019**, *19*, 1344–1351. [CrossRef] [PubMed]

33. Zielke, C.; Pan, C.W.; Gutierrez Ramirez, A.J.; Feit, C.; Dobson, C.; Davidson, C.; Sandel, B.; Abbyad, P. Microfluidic Platform for the Isolation of Cancer-Cell Subpopulations Based on Single-Cell Glycolysis. *Anal. Chem.* **2020**, *92*, 6949–6957. [CrossRef] [PubMed]

34. Stephan, K.; Pittet, P.; Renaud, L.; Kleimann, P.; Morin, P.; Ouaini, N.; Ferrigno, R. Fast prototyping using a dry film photoresist: Microfabrication of soft-lithography masters for microfluidic structures. *J. Micromech. Microeng.* **2007**, *17*, N69. [CrossRef]

35. Abramoff, M.D.; Magalhaes, P.J.; Ram, S.J. Image processing with ImageJ. *Biophotonics Int.* **2004**, *11*, 36–43.

36. Tran-Thi, T.H.; Gustavsson, T.; Prayer, C.; Pommeret, S.; Hynes, J.T. Primary ultrafast events preceding the photoinduced proton transfer from pyranine to water. *Chem. Phys. Lett.* **2000**, *329*, 421–430. [CrossRef]

37. Tolbert, L.M.; Solntsev, K.M. Excited-state proton transfer: From constrained systems to "super" photoacids to superfast proton transfer. *Acc. Chem. Res.* **2002**, *35*, 19–27. [CrossRef]

38. Wang, J.; Dong, J. Optical waveguides and integrated optical devices for medical diagnosis, health monitoring and light therapies. *Sensors* **2020**, *20*, 3981. [CrossRef]

39. Clement, N.R.; Gould, J.M. Pyranine (8-Hydroxy-1,3,6-pyrenetrisulfonate) as a Probe of Internal Aqueous Hydrogen Ion Concentration in Phospholipid Vesicles. *Biochemistry* **1981**, *20*, 1534–1538. [CrossRef]

40. Smith, K.K.; Kaufmann, K.J.; Huppert, D.; Gutman, M. Picosecond proton ejection: An ultrafast pH jump. *Chem. Phys. Lett.* **1979**, *64*, 522–527. [CrossRef]

41. Kotlyar, A.B.; Borovok, N.; Raviv, S.; Zimanyi, L.; Gutman, M. Fast redox perturbation of aqueous solution by photoexcitation of pyranine. *Photochem. Photobiol.* **1996**, *63*, 448–454. [CrossRef]

42. de Borba, E.B.; Amaral, C.L.C.; Politi, M.J.; Villalobos, R.; Baptista, M.S. Photophysical and photochemical properties of pyranine/methyl viologen complexes in solution and in supramolecular aggregates: A switchable complex. *Langmuir* **2000**, *16*, 5900–5907. [CrossRef]

43. Elmore, S. Apoptosis: A Review of Programmed Cell Death. *Toxicol. Pathol.* **2007**, *35*, 495–516. [CrossRef] [PubMed]
44. Clausell-Tormos, J.; Lieber, D.; Baret, J.C.; El-Harrak, A.; Miller, O.J.; Frenz, L.; Blouwolff, J.; Humphry, K.J.; Köster, S.; Duan, H.; et al. Droplet-Based Microfluidic Platforms for the Encapsulation and Screening of Mammalian Cells and Multicellular Organisms. *Chem. Biol.* **2008**, *15*, 427–437. [CrossRef]
45. Zivic, N.; Kuroishi, P.K.; Dumur, F.; Gigmes, D.; Dove, A.P.; Sardon, H. Recent Advances and Challenges in the Design of Organic Photoacid and Photobase Generators for Polymerizations. *Angew. Chem. Int. Ed.* **2019**, *58*, 10410–10422. [CrossRef]
46. Fradet, E.; McDougall, C.; Abbyad, P.; Dangla, R.; McGloin, D.; Baroud, C.N. Combining rails and anchors with laser forcing for selective manipulation within 2D droplet arrays. *Lab Chip* **2011**, *11*, 4228–4234. [CrossRef]

Publisher's Note: MDPI stays neutral with regard to jurisdictional claims in published maps and institutional affiliations.

Article

A Flexible, Microfluidic, Dispensing System for Screening Drug Combinations

Mark Davies [1,2,*], **Mannthalah Abubaker** [1] **and Lorraine Bible** [1,2]

[1] Bernal Institute, University of Limerick, Limerick V94 T9PX, Ireland; Mannthalah.Abubaker@ul.ie (M.A.); Lorraine.Bible@ul.com (L.B.)
[2] Hooke Bio Ltd., Clare V14 XH92, Ireland
*   Correspondence: Mark.Davies@ul.ie

Received: 25 September 2020; Accepted: 17 October 2020; Published: 18 October 2020

Abstract: It is known that in many cases a combination of drugs is more effective than single-drug treatments both for reducing toxicity and increasing efficacy. With the advent of organoid screens, personalised medicine has become possible for many diseases. Automated pipetting to well plates is the pharmaceutical industry standard for drug screening, but this is relatively expensive and slow. Here, a rotary microfluidic system is presented that can test all possible drug combinations at speed with the use of droplets. For large numbers of combinations, it is shown how the experimental scale is reduced by considering drug dilutions and machine learning. As an example, two cases are considered; the first is a three-ring and three radii configuration and the second is a four ring and forty-eight radii configuration. Between these two, all other cases are shown to be possible. The proposed commercial instrument is shown to be flexible, the user choosing which wells to fill and which driver-computational sub-routine to select. The major issues addressed here are the programming theory of the instrument and the reduction of droplets to be generated by drug dilutions and machine learning.

Keywords: dilutions; Microfluidics; drug combinations; screening; droplets

1. Introduction

Combinatorial drug therapy has shown promising results for treating complex diseases such as cancer, HIV, or malaria [1–3]. The administration of multiple compounds has shown a reduction in side effects and toxicity compared to monotherapy and lowers the risk of resistance development [4–8]. With drug combinations, a more targeted treatment known as personalized medicine becomes possible as different drugs can be tested on the cells of individual patients, thus customizing the care towards their specific response [9–13]. This will be particularly important with the use of 3D cell culture technologies such as organoids. Organoids can be established from patient-derived healthy and tumour tissue samples that self-organise in culture to acquire in-vivo like organ complexity that have been shown to be more translatable to in-vivo physiology [14–16]. However, to achieve this, large libraries consisting of many active ingredients must be screened to find the required response [17]. Currently, drug combinatorial tests are limited to combinations of two due to constraints in modern technology specific to drug screening [18]. The cost and time that is required by the existing technology make it unfeasible to test for all possible drug combinations [19,20]. Furthermore, to explore all the possible combinations leads to challenges associated with the 'Combinatorial Explosion' in which the number of combinations increases factorially with the number of drugs tested [21].

Figure 1 shows a prototype for ultra-high throughput, rotary, microfluidic drug screening. A novel method of dispensing aqueous-in-oil droplets, Gap Switch Technology (GST) is described in more detail in [22]. Stainless steel wells are filled from above, droplets are then dispensed from a 0.5 mm internal diameter outlet from below. Wells are designed to contain a range of biological fluids. Interfacial and capillary forces prevent fluid flow between dispensing. When a brass dispensing shuttle aligns with the well outlet, a liquid bridge is formed, and fluid flow is induced. The aqueous phase then encounters the immiscible carrier silicone oil, where the aqueous phase necks and is sheared off as the shuttle moves past the well. The aqueous phase is wrapped in the oil compartmentalising a droplet, Figure 2. This process is repeated to dispense and mix a second droplet. The droplets are efficiently mixed by having them of a different volume, hence the Laplace pressure difference drives mixing, vorticity. The mixed droplets are then transferred into a reusable polytetrafluoroethylene (PTFE) channel, for incubation and analysis.

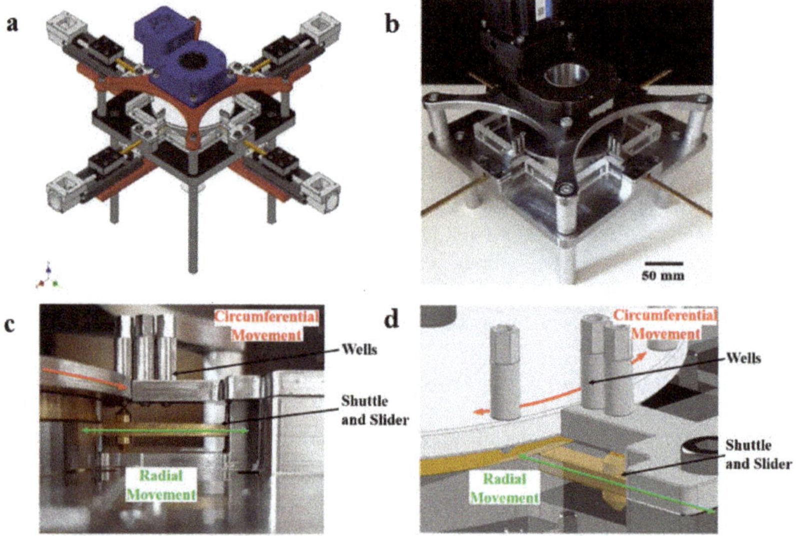

Figure 1. Prototype [22]. (**a**) CAD image of prototype. (**b**) Four-by-two prototype. (**c**) Circumferential movement of the ring is denoted by the red arrow and radial movement of the shuttle by the green arrow. (**d**) Corresponding CAD image.

An industry standard 96 well plate has a total assay volume in the order of 10^{-9} m^3. The mixed droplet volume is of order 6×10^{-11} m^3, making it significantly smaller than a well. There is no known reason why it should not be smaller. The aim is to provide a faster and lower-cost method of combinatorial drug screening. Operating costs are reduced by removing the reliance on pipettes and well plates and using smaller assay/reagent samples and removing the need for complex robotic set-ups required for current drug screening technologies [23]. Dispensing time is reduced due to the instruments enhanced mixing and encapsulation. These mass transport properties are due to the high surface area to volume ratio, facilitating fast and controlled heating of droplets and creating shorter diffusion distances between molecules [24–28].

The GST has been designed to be scaled into a rotational design, Figure 1, to increase microfluidic throughput. The design allows movement of wells by concentric rotating rings into the pathway of radial shuttle movements. This improves the throughput of the system through parallelization [29]. The shuttle can then selectively dispense and mix drugs from these wells along the radial shuttle pathways. The flexibility of the design is due to the many potential configurations, depending on which wells are filled and which computational sub-routine is selected. Four instrument strategies are discussed:

1. Measure every combination without dilutions.
2. Measure every meaningful combination with dilutions.
3. For large numbers of droplets, measure 1 and 2 with machine learning.
4. For 3 measure single dilutions first, then next dilution until synergy is found.

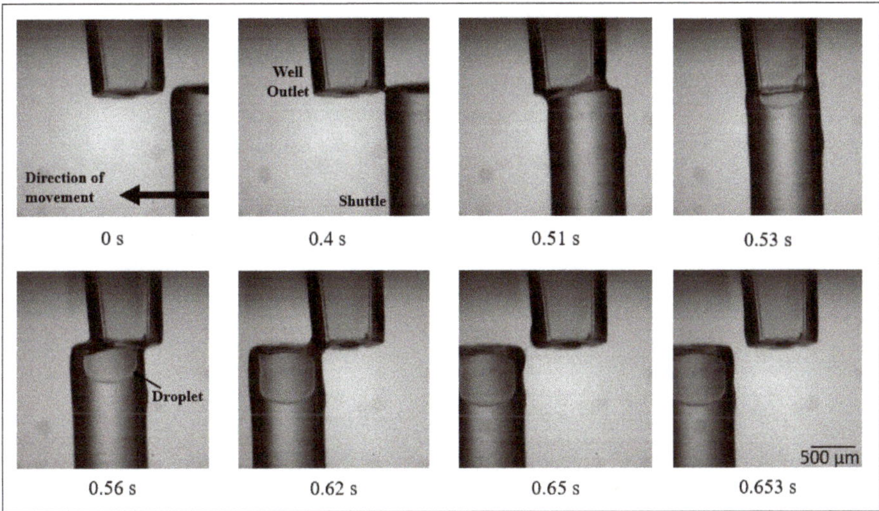

Figure 2. Gap Switch Technology (GST), Droplet Generation and Mixing. High-speed camera images taken of the GST. Images were taken using an IDT X-Stream XS-4 CMOS Camera at 1000 frames per second. The droplets are completely surrounded by silicone oil and therefore never contact a solid surface. This eliminates carry-over contamination and makes the system continuously reusable. Shuttle velocity is approximately 4mm/s. The dispensed droplet can be seen in the lower tube. The tubing is PTFE which attracts the oil and repels the aqueous, hence the wrapped droplets.

2. Three-by-Three Example

To introduce the system a three-by-three configuration is considered, nomenclature is given in Figure 3. Figure 4 shows a three radii, three ring build. It operates through three types of movements;

1. Circumferential movements: movement from one well to another well on the same ring,
2. Radial movements: movement from a well on one ring to a well on a different ring,
3. Mixed movements: a combination of radial and circumferential movements.

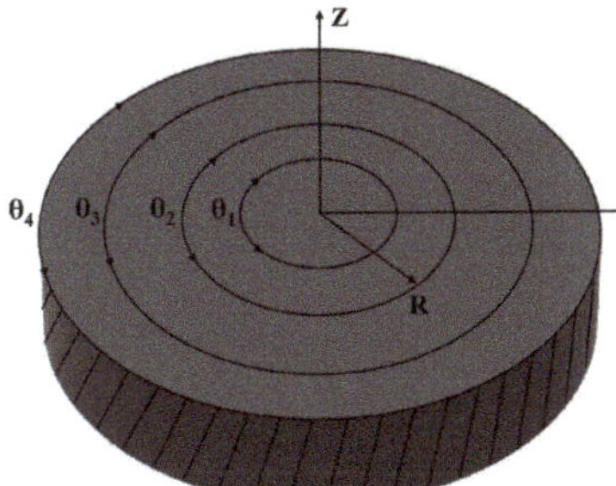

Figure 3. Nomenclature.

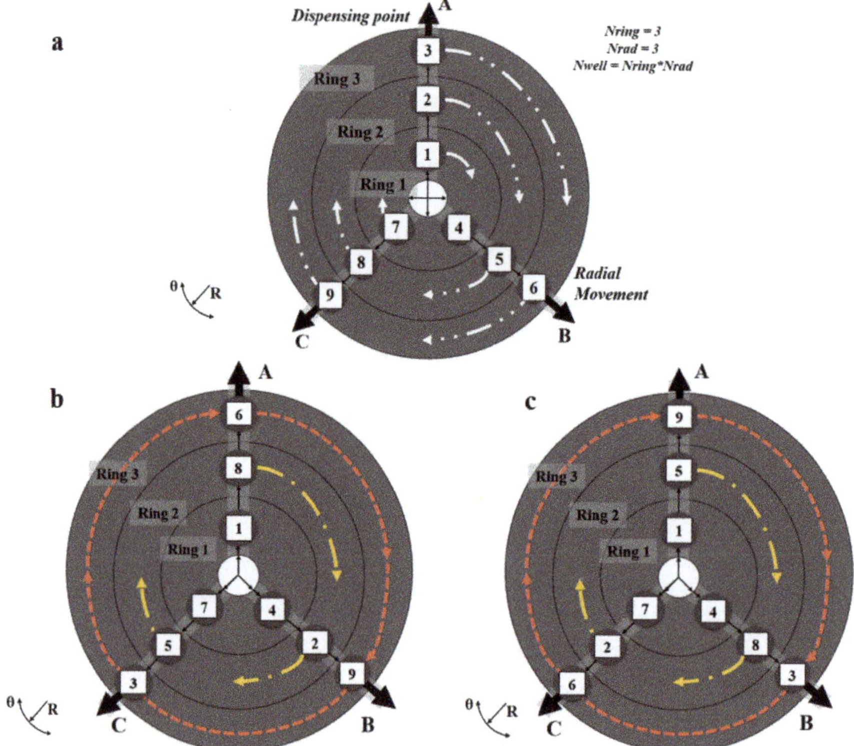

Figure 4. Three-by-Three Configuration. (**a**) Overview of a 3 × 3 configuration. Drug combinations occur through either circumferential movement, the movement of a droplet from a well on one radius to a well on another radius, or through radial movements, the movement of a droplet along the same radius. There is a total of 84 combinations, requiring 28 positions for all combinations. (**b**) shows rotor position after movement. The movement of a droplet on ring 2 from a well on radius 1 to a well on radius 2 (a. 120° rotation) and a droplet on ring 3 moving from a well on radius 1 to a well on radius 3 (a 240° rotation). (**c**) shows the next positions following (**b**). A, B and C are the outlet lines.

There are nine filled wells containing either drugs, cells or organoids. The figure shows how the different wells, rings and radii interact both mechanically and fluid mechanically. The shuttles move radially to collect droplets from wells on the same radius and the rings would move circumferentially to allow the shuttle to collect droplets from wells on different radii. The shuttle can pass under a well without taking a droplet if it is moving at a high enough velocity. If drugs are combined in sets of three, there will be a total of eighty-four combinations, Equation (1). This requires the instrument to take up twenty-eight positions. The mathematics of the instrument is explained in the next section. For commercialisation, a maximum forty-eight by four configuration is proposed. This is twice the number of wells on a ninety-six well plate which is considered the industry standard [30].

3. Discrete Analysis

The binomial equation [31],

$$_nC_r = \frac{n!}{r!(n-r)!} \tag{1}$$

applied to the instrument design, Figures 3 and 4,

$$N_p \, N_{rad} \; = \; \frac{\left(N_{ring} \, N_{rad}\right)!}{r! \left(N_{ring} \, N_{rad} \, - \, r\right)!} \tag{2}$$

describes its operation without dilutions. The number of droplets to complete a test is determined by knowing the number of radii, the configuration and the number of combinations to test. The number of droplets,

$$_nC_r \; = \; N_p \, N_{rad} \tag{3}$$

maybe derived in two, depending on how they are formed,

$$_nC_r \; = \; _nC_{rr} \, + \, _nC_{rc} \tag{4}$$

If dilutions are used,

$$_nC_r' = \, _nC_{r_r}' + \, _nC_{r_c}' \tag{5}$$

in which case,

$$_nC_r > \, _nC_r' \tag{6}$$

which always gives a throughput advantage.

With dilutions the following restriction is applied,

$$N_{ring} \; = \; N_{dil} \; = \; r \tag{7}$$

then the required number of droplets is,

$$_nC_r' = r \, N_{rad}^{\,r-1} \, \left(N_{rad} \, - \, r \, + \, 1\right) \tag{8}$$

by inspection, which cannot be proven, only shown. All of $_nC_r$ and $_nC_r{}'$ are integers. The inequality, (6), becomes more numerically advantageous as the number of combinations increases, Table 1. In the simple example of a three by three configuration, Figure 4, the number of droplets needed is reduced from eighty-four to twenty-seven. The addition of four components, three drugs and a cell or multicell line, is shown in Figure 5 for a four radii configuration.

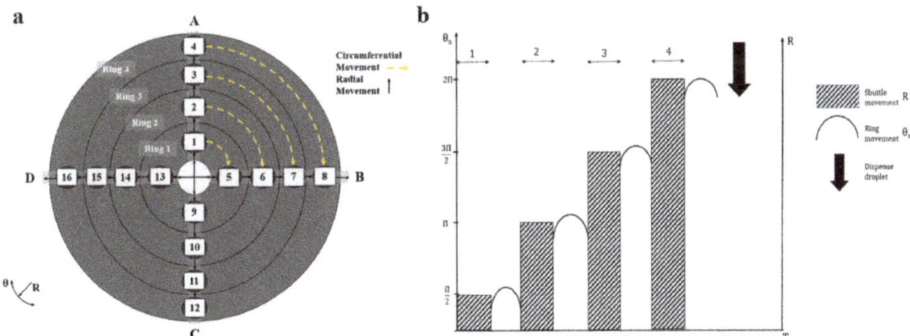

Figure 5. Movement on a four-by-four configuration. (**a**) 4 × 4 configuration. (**b**) illustrates the movement of the shuttles and rings. The shuttle and rings move in programable increments. It begins with the shuttle moving to the first ring to collect a droplet, the ring then rotates clockwise towards the next radii to collect the next droplet. These movements are performed until all the droplets are produced and the combinations dispensed.

Table 1. Different configurations for the proposed 48 × 4 commercial design. $_nC_r$ and $_nC_r{'}$ for different builds and combinations are given. The number of combinations r corresponds to the number of rings, N_{ring}. Each radius has a different drug and each ring a different dilution. The effect of dilution can be observed in this table. The effect of machine learning can also be seen by comparing the time taken for the droplets when dilutions are considered to the time taken with machine learning. All configurations are for the same instrument.

r	N_{rad}	$_nC_r$	$_nC_r{'}$	$N_c{'}$	$N_r{'}$	$N_p{'}$	Time (days)	Time ML (days), Single Ring, See ML Section
2	2	6	4	2	2	2	0.0002	0.000001
2	3	15	12	6	6	4	0.0004	0.000003
2	4	28	24	12	12	6	0.0007	0.000007
2	6	66	60	30	30	10	0.001	0.00002
2	8	120	112	56	56	14	0.002	0.00003
2	10	190	180	90	90	18	0.002	0.00005
2	20	780	760	380	380	38	0.004	0.0002
2	30	1770	1740	870	870	58	0.007	0.0005
2	48	4560	4512	2256	2256	94	0.01	0.0014
3	3	84	27	3	24	9	0.001	0.000002
3	4	220	96	12	84	24	0.003	0.000007
3	6	816	432	60	372	72	0.008	0.00004
3	8	2024	1152	168	984	144	0.02	0.0001
3	10	4060	2400	360	2040	240	0.03	0.0002
3	20	34,220	21,600	3420	18,180	1080	0.13	0.0021
3	30	117,480	75,600	12,180	63,420	2520	0.3	0.0073
3	40	280,840	182,400	29,640	152,760	4560	0.5	0.0179
3	48	487,344	317,952	51,888	266,064	6624	0.5	0.0313
4	4	1820	256	4	252	64	0.007	0.000002
4	6	10,626	2592	60	2532	432	0.05	0.00003
4	8	35,960	10,240	280	9960	1280	0.15	0.0002
4	10	91,390	28,000	840	27,160	2800	0.3	0.0005
4	20	1,581,580	544,000	19,380	524,620	27,200	3.15	0.0117
4	30	8,214,570	2,916,000	109,620	2,806,380	97,200	11.3	0.0661
4	35	15,329,615	5,488,000	209,440	5,278,560	156,800	18.1	0.1263
4	40	26,294,360	9,472,000	365,560	91,06440	236,800	27.4	0.2204
4	48	54,870,480	19,906,560	778,320	19,128,240	414,720	48	0.4692

4. Reducing the Number of Combinations by Dilutions

Taking different dilutions of the drugs is important as it not only allows for a range of concentrations to be tested, but also reduces interference from other substances that may be present in the samples [32]. Therefore, dilutions are often a necessary part of any drug testing experiment. In this context, dilutions can also be used to reduce the combinatorial explosions. For a forty-eight-by-four system, this offers the user a maximum input of forty-eight different drugs, in dilutions of four. Using Equation (1), a total of some fifty-four million droplets need be produced to complete all possible combinations. Every droplet will need to be repeated several times, so the maximum number of droplets is over two-hundred million. Hence the need to reduce it.

With dilutions, many of the droplets contain redundant samples, combinations that contain two or more dilutions of the same drugs. These samples are not needed. Therefore, by taking account of dilutions, the number of total droplets to be produced for the test run can be significantly reduced, Equation (8). Figure 6 illustrates the reduction in droplets between $_nC_r$ and $_nC_r{'}$. For a forty-eight-by-four system, with $r = 4$, the number of droplets reduces from around fifty million to around nineteen million.

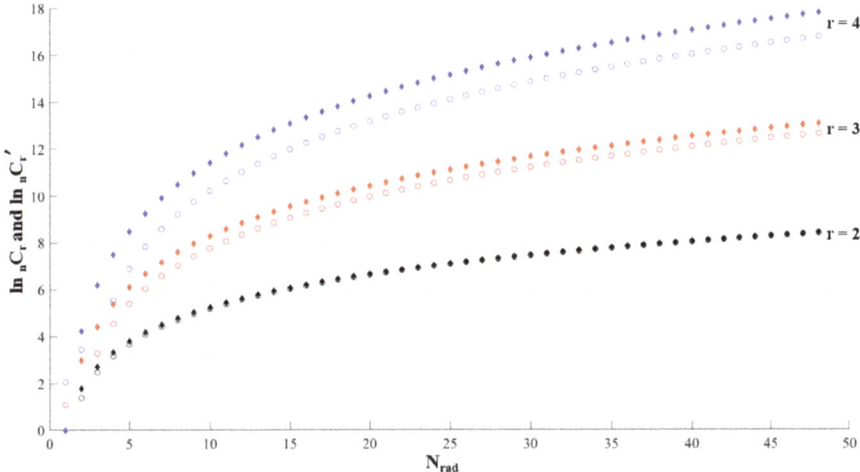

Figure 6. The logarithmic difference between Equation (1) and Equation (8). The circular points are $_nC_r$ and the diamonds $_nC_r{'}$. As r increases so does the difference between $_nC_r$ and $_nC_r{'}$.

5. Commercial Instrument

The proposed commercial design has a maximum of forty-eight drugs in combinations of four, Figure 7. With no dilutions, this has of order of fifty-five million droplets required and over a million rotor movements to dispense them. There are many lesser configurations, depending on which wells are filled and which algorithm subroutine is selected.

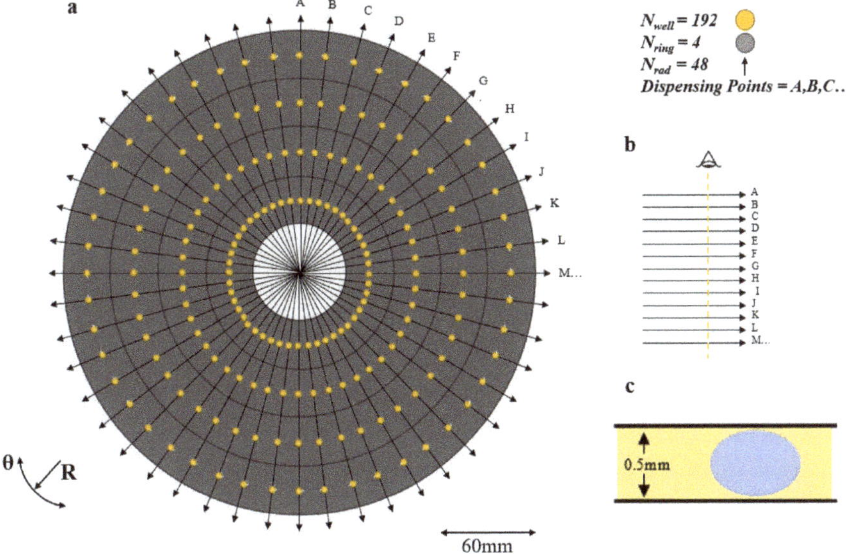

Figure 7. Proposed Commercial 48×4 Configuration. (**a**) 48×4 system containing forty-eight drugs in four different dilutions. Each radius of the system contains a single drug, with each ring in that radius containing a dilution of that drug. Each dot represents a well containing a drug dilution. Any number of radii and rings may be utilised depending on the needs of the user. At maximum some produces 53 million combinations of four. (**b**) After exiting the rotors, the droplets are read using either fiber optics or laser LEDs. (**c**) Typical droplet completely surrounded by silicone oil.

If the maximum is chosen with dilutions, there are then forty-eight different drugs and four dilutions of each drug. Equation (4) gives 19,906,560 combinations and 414,730 positions to dispense them. This requires of order forty-eight days plus the well refill time. In personalized medicine, this is a limitation on the use of multiple drug combinatorial discovery with organoids. For these reasons machine learning is proposed.

6. Machine Learning

Machine learning (ML) uses algorithms to process data, learns from it, and makes a prediction about the future state of any new data set [33]. With ML, a system can be trained to be able to predict the outcome of a task by using a large amount of data and algorithms. In the context of combinatorial drug screening, ML has the potential to reduce the scale by speeding up testing. This is particularly beneficial for personalised medicine as accurate predictions to tailor patient-specific treatments would be faster. ML can be used to focus in on a specific combination when synergy is detected using the Loewe additivity or Bliss model for example [18]. When a ML test is run, a circumferential movement on a single ring would be step one. If synergy were to be found between three drugs, then the system will focus on that region (the drugs in that area). This can be achieved through scoring. If the drugs display additive interactions, in which there is a difference in the way they interact together, they would be scored 1. If they display antagonistic interactions, in which they hinder the response, they would be scored 0, and if they show synergistic interactions, they score 2. This is one method of synergy-searching. After one circumference is tested with no synergy, step two would be to test another, etc. This may be a more hands-on process of filling only the wells in a circumference until synergy is found.

7. Conclusions

It is demonstrated how the automatic microfluidic design has great flexibility in the many configurations that may be chosen by the user. A 192 well system is proposed which can produce over 50 million combinations for synergy-searching. It is shown how this number is reduced by taking dilutions, by machine learning, and by more user interaction with the instrument. Throughput and drug volume consumed may follow a type of Moore's law in the future.

8. Patents

Davies, M. Microfluidic device with channel plates. WO 2015/173651; EP3142790, 20 March 2019.

Davies, M; Galvin, M. Rotating mechanical device for generating a combinatorial droplet library. WO 2016/207721; EP3314046, 24 April 2019.

Author Contributions: M.D. devised the project, the main conceptual ideas, technical detail and theory. M.A. assisted M.D in performing calculations, verified numerical results and created the figures and wrote the manuscript. L.B. provided instrument images and assisted in editing of the manuscript. All authors contributed to the final manuscript. All authors have read and agreed to the published version of the manuscript

Funding: This project was supported by financial assistance via Enterprise Ireland's Commercialisation Funds [CF 20150176 and CF 20133342] and the Disruptive Technologies Innovation Fund (DTIF) under Project 2040 [164610/RR], which partially funded this research.

Conflicts of Interest: The authors declare no conflict of interest.

Nomenclature

Symbol	Description
N_c	Number of circumferential movements
$N_{c'}$	Number of circumferential movements with dilutions
N_{dil}	Number of dilutions
N_r	Number of radial movements
$N_{r'}$	Number of radial movements with dilutions

N_{rad}	Number of radii of the configuration
N_{ring}	Number of rings of the configuration
N_p	Number of positions the system must perform
N_{well}	Number of wells
$_mC_{n_R}$	Number of droplets formed radially
$_mC'_{n_R}$	Number of droplets formed radially with dilutions
$_mC_{n_C}$	Number of droplets formed circumferentially
$_mC'_{n_C}$	Number of droplets formed circumferentially with dilutions
$_nC_r$	Number of droplets without dilutions
$_nC_r{}'$	Number of droplets with dilutions
n	Number of wells on the system
r	Number of combinations
z	Shuttle dispensing direction
R	Radial movements, radial direction
θ	Angular movements

References

1. Goutsouliak, K.; Veeraraghavan, J.; Sethunath, V.; De Angelis, C.; Osborne, C.K.; Rimawi, M.F.; Schiff, R. Towards Personalized Treatment for Early Stage HER2-Positive Breast Cancer. *Nat. Rev. Clin. Oncol.* **2019**, *17*, 233–250. [CrossRef] [PubMed]
2. Freedberg, K.A.; Losina, E.; Weinstein, M.C.; Paltiel, A.D.; Cohen, C.J.; Seage, G.R.; Craven, D.E.; Zhang, H.; Kimmel, A.D.; Goldie, S.J. The Cost Effectiveness of Combination Antiretroviral Therapy for HIV Disease. *N. Engl. J. Med.* **2001**, *344*, 824–831. [CrossRef] [PubMed]
3. Alven, S.; Aderibigbe, B. Combination Therapy Strategies for the Treatment of Malaria. *Molecules* **2019**, *24*, 3601. [CrossRef] [PubMed]
4. Folkesson, E.; Niederdorfer, B.; Nakstad, V.T.; Thommesen, L.; Klinkenberg, G.; Lægreid, A.; Flobak, Å. High-Throughput Screening Reveals Higher Synergistic Effect of MEK Inhibitor Combinations in Colon Cancer Spheroids. *Sci. Rep.* **2020**, *10*, 11574. [CrossRef] [PubMed]
5. Neutel, J. Advantages of Combination Therapy Compared with Monotherapy. In *Combination Therapy in Hypertension*; Springer Healthcare: Tarporley, UK, 2011; pp. 23–35. [CrossRef]
6. Schmid, A.; Wolfensberger, A.; Nemeth, J.; Schreiber, P.W.; Sax, H.; Kuster, S.P. Monotherapy versus Combination Therapy for Multidrug-Resistant Gram-Negative Infections: Systematic Review and Meta-Analysis. *Sci. Rep.* **2019**, *9*, 1–11. [CrossRef]
7. Sun, W.; Sanderson, P.; Zheng, W. Drug Combination Therapy Increases Successful Drug Repositioning. *Drug Discov. Today* **2016**, *21*, 1189–1195. [CrossRef]
8. Clarke, P.A.; Roe, T.; Swabey, K.; Hobbs, S.M.; McAndrew, C.; Tomlin, K.; Westwood, I.; Burke, R.; van Montfort, R.; Workman, P. Dissecting Mechanisms of Resistance to Targeted Drug Combination Therapy in Human Colorectal Cancer. *Oncogene* **2019**, *38*, 5076–5090. [CrossRef] [PubMed]
9. Chan, I.S.; Ginsburg, G.S. Personalized Medicine: Progress and Promise. *Annu. Rev. Genom. Hum. Genet.* **2011**, *12*, 217–244. [CrossRef] [PubMed]
10. Gorshkov, K.; Chen, C.Z.; Marshall, R.E.; Mihatov, N.; Choi, Y.; Nguyen, D.-T.; Southall, N.; Chen, K.G.; Park, J.K.; Zheng, W. Advancing Precision Medicine with Personalized Drug Screening. *Drug Discov. Today* **2019**, *241*, 272–278. [CrossRef]
11. Schork, N.J. Personalized Medicine: Time for One-Person Trials. *Nature* **2015**, *520*, 609–611. [CrossRef]
12. Redekop, W.K.; Mladsi, D. The Faces of Personalized Medicine: A Framework for Understanding Its Meaning and Scope. *Value Health* **2013**, *16*, S4–S9. [CrossRef]
13. Weatherall, D.; Bodmer, W.; Breckenridge, A.K.; Davies, D.; Pirmohamed, M.; Wilkie, A. Recent Developments in Personalised Medicine, Royal Society. Available online: https://royalsociety.org/-/media/policy/Publications/2015/personalised-medicine-report.pdf (accessed on 17 March 2020).
14. Fang, Y.; Eglen, R.M. Three-Dimensional Cell Cultures in Drug Discovery and Development. *Slas Discov. Adv. Life Sci. R D* **2017**, *22*, 456–472. [CrossRef]

15. Lehmann, R.; Lee, C.M.; Shugart, E.C.; Benedetti, M.; Charo, R.A.; Gartner, Z.; Hogan, B.; Knoblich, J.; Nelson, C.M.; Wilson, K.M. Human Organoids: A New Dimension in Cell Biology. *Mol. Biol. Cell* **2019**, *30*, 1129–1137. [CrossRef] [PubMed]

16. Takebe, T.; Wells, J.M. Organoids by Design. *Science* **2019**, *364*, 956–959. [CrossRef] [PubMed]

17. Feala, J.D.; Cortes, J.; Duxbury, P.M.; Piermarocchi, C.; McCulloch, A.D.; Paternostro, G. Systems Approaches and Algorithms for Discovery of Combinatorial Therapies. *Wiley Interdiscip. Rev. Syst. Biol. Med.* **2010**, *2*, 181–193. [CrossRef]

18. Pemovska, T.; Bigenzahn, J.W.; Superti-Furga, G. Recent Advances in Combinatorial Drug Screening and Synergy Scoring. *Curr. Opin. Pharmacol.* **2018**, *42*, 102–110. [CrossRef]

19. Sun, J.; Warden, A.R.; Ding, X. Recent Advances in Microfluidics for Drug Screening. *Biomicrofluidics* **2019**, *13*, 061503. [CrossRef]

20. Sinha, S.; Vohora, D. Drug Discovery and Development. In *Pharmaceutical Medicine and Translational Clinical Research*; Academic Press: New York, NY, USA, 2018; pp. 19–32. [CrossRef]

21. Velleman, D.J. Exponential vs. Factorial. *Am. Math. Mon.* **2006**, *113*, 689. [CrossRef]

22. Cliffe, F.E.; Lyons, M.; Murphy, D.C.; McInerney, L.; Hurley, N.; Galvin, M.A.; Mulqueen, J.; Bible, L.B.; Marella, C.; Kelleher, M.; et al. Droplet Combinations: A Scalable Microfluidic Platform for Biochemical Assays. *SLAS Technol. Transl. Life Sci. Innov.* **2019**, *25*, 140–150. [CrossRef]

23. Wildey, M.J.; Haunso, A.; Tudor, M.; Webb, M.; Connick, J.H. High-Throughput Screening. *Annu. Rep. Med. Chem.* **2017**, *50*, 149–195. [CrossRef]

24. Clausell-Tormos, J.; Lieber, D.; Baret, J.-C.; El-Harrak, A.; Miller, O.J.; Frenz, L.; Blouwolff, J.; Humphry, K.J.; Köster, S.; Duan, H.; et al. Droplet-Based Microfluidic Platforms for the Encapsulation and Screening of Mammalian Cells and Multicellular Organisms. *Chem. Biol.* **2008**, *15*, 427–437. [CrossRef]

25. Damiati, S.; Kompella, U.B.; Damiati, S.A.; Kodzius, R. Microfluidic Devices for Drug Delivery Systems and Drug Screening. *Genes* **2018**, *9*, 103. [CrossRef]

26. Dittrich, P.S.; Manz, A. Lab-on-a-Chip: Microfluidics in Drug Discovery. *Nat. Rev. Drug Discov.* **2006**, *5*, 210–218. [CrossRef]

27. Huang, H.; Yu, Y.; Hu, Y.; He, X.; Usta, O.B.; Yarmush, M.L. Generation and Manipulation of Hydrogel Microcapsules by Droplet-Based Microfluidics for Mammalian Cell Culture. *Lab Chip* **2017**, *17*, 1913–1932. [CrossRef]

28. Wang, Y.; Chen, Z.; Bian, F.; Shang, L.; Zhu, K.; Zhao, Y. Advances of Droplet-Based Microfluidics in Drug Discovery. *Expert Opin. Drug Discov.* **2020**, *15*, 969–979. [CrossRef] [PubMed]

29. Yadavali, S.; Lee, D.; Issadore, D. Robust Microfabrication of Highly Parallelized Three-Dimensional Microfluidics on Silicon. *Sci. Rep.* **2019**, *9*, 12213. [CrossRef] [PubMed]

30. Lee, P.J.; Ghorashian, N.; Gaige, T.A.; Hung, P.J. Microfluidic System for Automated Cell-Based Assays. *JALA* **2007**, *12*, 363–367. [CrossRef] [PubMed]

31. Boyer, C.B.; Merzbach, U.C. A History of Mathematics. *Biometrics* **1993**, *49*, 674. [CrossRef]

32. Ben-David, A.; Davidson, C.E. Estimation Method for Serial Dilution Experiments. *J. Microbiol. Methods* **2014**, *107*, 214–221. [CrossRef] [PubMed]

33. Vamathevan, J.; Clark, D.; Czodrowski, P.; Dunham, I.; Ferran, E.; Lee, G.; Li, B.; Madabhushi, A.; Shah, P.; Spitzer, M.; et al. Applications of Machine Learning in Drug Discovery and Development. *Nat. Rev. Drug Discov.* **2019**, *18*, 463–477. [CrossRef] [PubMed]

Publisher's Note: MDPI stays neutral with regard to jurisdictional claims in published maps and institutional affiliations.

 *micromachines*

Article

Programmable µChopper Device with On-Chip Droplet Mergers for Continuous Assay Calibration

Nan Shi and Christopher J. Easley *

Department of Chemistry and Biochemistry, Auburn University, Auburn, AL 36849, USA; nzs0063@auburn.edu
* Correspondence: chris.easley@auburn.edu

Received: 5 June 2020; Accepted: 22 June 2020; Published: 25 June 2020

Abstract: While droplet-based microfluidics is a powerful technique with transformative applications, most devices are passively operated and thus have limited real-time control over droplet contents. In this report, an automated droplet-based microfluidic device with pneumatic pumps and salt water electrodes was developed to generate and coalesce up to six aqueous-in-oil droplets (2.77 nL each). Custom control software combined six droplets drawn from any of four inlet reservoirs. Using our µChopper method for lock-in fluorescence detection, we first accomplished continuous linear calibration and quantified an unknown sample. Analyte-independent signal drifts and even an abrupt decrease in excitation light intensity were corrected in real-time. The system was then validated with homogeneous insulin immunoassays that showed a nonlinear response. On-chip droplet merging with antibody-oligonucleotide (Ab-oligo) probes, insulin standards, and buffer permitted the real-time calibration and correction of large signal drifts. Full calibrations (LOD_{conc} = 2 ng mL^{-1} = 300 pM; LOD_{amt} = 5 amol) required <1 min with merely 13.85 nL of Ab-oligo reagents, giving cost-savings 160-fold over the standard well-plate format while also automating the workflow. This proof-of-concept device—effectively a microfluidic digital-to-analog converter—is readily scalable to more droplets, and it is well-suited for the real-time automation of bioassays that call for expensive reagents.

Keywords: droplets; lock-in detection; real-time calibration; homogeneous immunoassay; on-chip mergers; pneumatic valves; programmable droplet formation

1. Introduction

Droplet-based microfluidics is an important subcategory of microfluidic technology. In these types of micro-devices, small droplets are generated and viewed as individual reactors, and they provide powerful platforms for confining samples to small volumes for subsequent manipulation, reaction, and analysis [1]. In the last decade, droplet microfluidics has been widely used in a broad range of biochemical fields, such as nucleic acid/molecule analysis [2,3], drug delivery [4], cell-to-cell communication [5], cell screening [6], tissue analysis [7–9], and so on. To ensure constant and predictable outcomes in these applications, it is essential to generate highly uniform droplet volumes [10–12], and researchers have developed various methods to do so.

Microfluidic droplet formation techniques can be divided into two categories: passive and active. High throughput droplet generation is much simpler and faster to achieve with passive methods, an obvious advantage in applications that require enormous experimental throughput [13]. By contrast, a major benefit of active droplet generation is its higher flexibility in droplet volume and production rate [14]. Because the vast majority of biochemical reactions and analyses require multiplexed reagents, multiple timed steps, and often multiple conditions (temperature, pH, ionic strength, etc.), tools that allow for a precise control of droplets on demand are becoming increasingly important. Significant efforts have been focused on active droplet formation using various approaches such as electric, magnetic, thermal, and mechanical control [15–18]. Considering the exquisite level of control that they provide, on-chip pneumatic valves [19] have been demonstrated as important players that provide an active, programmable droplet generation with high precision [7,9,15,20–22].

To improve programmability and precision, our laboratory has moved from passive droplet formation [11,12], to active fluidic resistors [21], to the gating of fluids with single pneumatic valves [8,22], and finally to on-chip valve-based pumps [7,9]. During this time, we revealed one less obvious benefit of active control: the ability to precisely control the frequency and phase of droplets, lock in the photodetector to that signal, and greatly reduce the detection limits—an approach we refer to as the μChopper [8,12,22]. With a control bandwidth of ±0.04 Hz using gating valves, the fluorescence detection limits were reduced more than 50-fold using simple microscope detection optics, and even single-cell fatty acid uptake was quantifiable in droplets [8]. An improved iteration of the μChopper with six aqueous input channels enabled several analytical modes to be programmed automatically, such as real-time continuous calibration, standard addition, and a mixed mode [22]. Despite these benefits, there remains a drawback with respect to the workflow in this type of microsystem. Reagents for multi-step or timed reactions must be manually pre-mixed and transported to the input micro-reservoirs, increasing the bench time and potential operator errors. The logical step is to add on-chip reagent mixing or to incorporate programmable droplet mergers.

The Ismagilov group and others have successfully initiated the mixing of reagents at the droplet forming structure [7,23–25], which can start reactions at a predictable position and provide control over timing. However, several issues limit the accuracy and preclude the universal application of this approach. First, inconsistent flow rates of solutions from individual aqueous channels can lead to fluctuating reagent volume ratios and significantly affect assay outcomes. Second, it is difficult to precisely and arbitrarily change the volume ratio of reagents, meaning that new channel designs will be needed for even minor adjustments. Several techniques to coalesce neighboring droplets were introduced to avoid these issues, such as hydrodynamic, magnetic, electric, or acoustic coalescence [26–30]. Among these, electrocoalescence has been the most widely used in droplet microfluidics by merging adjacent droplets with an alternating current (AC) electric field applied to nearby electrodes on the device. The development of in-channel "salt water electrodes" by the Abate group, where high-concentration salts can replace metal solder, has made this approach even more accessible [28].

Considering the benefits of pneumatically controlled droplet generation and electrocoalescence, here we have integrated our μChopper approach with active valve-based pumps and salt-water electrodes for the first time. This approach permits the fully automated, on-demand production and merging of several types of droplets in a programmable way. In this proof-of-concept work, we apply the device to the real-time, continuous calibration of fluorescent labels, and then we validate the system for the continuous calibration of a homogeneous insulin immunoassay that exhibits a nonlinear response. With the significant savings in reagent use, assay cost, and user time that are incurred, this device provides a novel means to carry out economical measurements with precious reagents in a static or real-time manner.

2. Materials and Methods

2.1. Materials and Equipment

All materials and equipment were obtained from sources within the USA. Buffers were prepared using deionized water filtered with a Barnstead MicroPure Water Purification system (ThermoFisher Scientific, Waltham, MA, USA). Citric acid, sodium phosphate, and sodium chloride were obtained from Millipore Sigma (Burlington, MA, USA). Polydimethylsiloxane (PDMS) precursors, SYLGARD 184 silicone elastomer base, and curing agent were purchased from Dow Corning (Midland, MI, USA). The silicon wafers were acquired from the Polishing Corporation of America (Santa Clara, CA, USA). Negative photoresists (SU-8 2015 and SU-8 2050) and SU-8 developer were purchased from MicroChem

(Westborough, MA, USA). Positive photoresist (AZ 40XT-11D) and AZ 300 MIF developer were obtained from AZ Electronic Materials USA (Somerville, NJ, USA). Fluorescein was purchased from Alfa Aesar (Ward Hill, MA, USA), and bovine serum albumin (BSA) was purchased from VWR (West Chester, PA, USA). Human Insulin FRET-PINCER Assay Kits were obtained from Mediomics, LLC (Saint Louis, MO, USA). Pico-Surf (2% in Novec 7500 oil), a perfluorocarbon surfactant, was purchased from Dolomite Microfluidics (Norwell, MA, USA) for stabilizing droplets against unwanted coalescence and to provide biocompatible surfaces within the droplets. Novec 7500 Engineered Fluid (HFE 7500) was acquired from 3M (St. Paul, MN, USA).

A high voltage amplifier (Model 2220) was purchased from Trek, Inc. (Lockport, NY, USA) and used for droplet merging. Fluorescence excitation and emission were accomplished using a Nikon Ti-E inverted fluorescence microscope (40X objective, 0.75 NA; Nikon Instruments Inc., Melville, NY, USA) interfaced to a CCD camera (CoolSnap HQ2; Photometrics Scientific, Tucson, AZ, USA). Fluorescence images were acquired by focusing on a chosen region of interest in the incubation channel (Figure 1A) and collecting at 100 frames s^{-1} through the green fluorescence filter cube (λ_{ex} = 470 ± 20 nm, λ_{em} = 525 ± 25 nm).

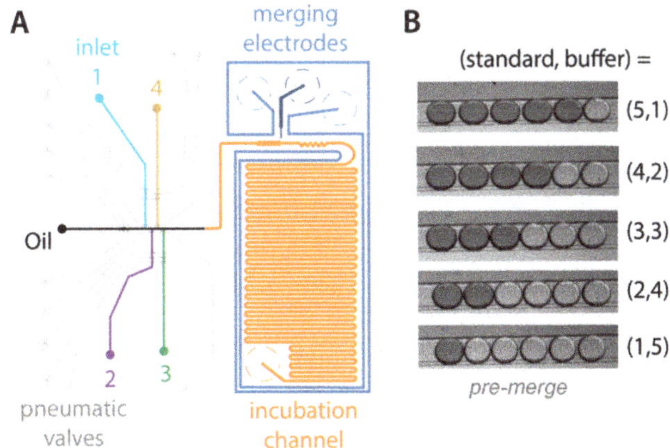

Figure 1. Microdevice design and operation. (**A**) Inlet aqueous reservoirs (1–4, colored) and one oil reservoir (black) were sampled by computer-controlled pumps based on pneumatic valves (light gray). Merging electrodes (dark blue) facilitated droplet coalescence in the widened merging region (orange), merged droplets were mixed in a zig-zag channel, and then assays were incubated in a long delay channel (orange) if needed prior to optical detection. (**B**) In this example, five ratios of standard mimics (dark) and buffer (transparent) were programmed on demand, then merged downstream. Images show the droplet groups prior to merging (see Videos S1 and S2).

2.2. Microfluidic Master Wafer Fabrication

Two master wafers for templating liquid channels and pneumatic control channels were fabricated using standard photolithography as described previously [9]. Channel layouts were designed in Adobe Illustrator software, and plastic film photomasks were printed at Fineline Imaging (Colorado Springs, CO, USA) at a 50,800 dpi resolution. For the pneumatic control channel layer, a ~20 μm layer of SU-2015 was spin-coated on the silicon wafer, which had been washed by 1 M H_2SO_4 and water in advance. The wafer was soft-baked at 95 °C or 5 min, after which ultraviolet (UV) light exposure for 2 min was accomplished on an in-house built UV lithography light exposure unit [31]. Finally, the wafer was developed for 5 min in SU-8 developer solution after a 5-min hard bake on a hot plate at 115 °C. The fluidic layer wafer was fabricated in a two-step protocol with both negative and positive photoresists, respectively. First, a 60-μm layer of SU-8 2050 was spun onto the pretreated silicon wafer, and the wafer was soft baked at 95 °C for 7 min. UV exposure with the first photomask was carried out for 90 s, hard baking was applied at 95 °C for 6 min, and then the SU-8 developer was applied. Next, a 40-μm layer of AZ 40 XT was spun onto the wafer at room temperature. The wafer was baked at 115 °C for 5 min, cooled to room temperature passively, and the second photomask was aligned over the wafer then exposed to UV light. A final hard bake at 115 °C was applied for 1.5 min, AZ developer was applied, and the AZ portion of the wafer was annealed to allow templating of rounded channel cross-sections by baking at 120 °C for 10 min.

2.3. Microchip Fabrication

After degassing under vacuum, 36 g of well-mixed PDMS precursor mixture (5:1 ratio, monomer: curing agent) was poured onto the flow channel patterned silicon wafer in an aluminum foil boat. Again after degassing, 5.12 g of PDMS precursor mixture (15:1 ratio, monomer:curing agent) was spin-coated onto the control layer at 2100 rpm for 45 s, creating a layer of ~40 μm thickness. Both the fluid layer and control layer were baked at 65 °C for 30 min in an oven. The flow channel layer was then cut to shape, then aligned and mated to the valve channel layer. The two mated layers were baked in an oven at 65 °C overnight to facilitate permanent bonding. The PDMS was peeled from the wafer, diced into individual devices, access reservoirs were punched, and the surfaces were washed with methanol and dried with N_2 gas. Each device was then irreversibly bonded to a glass slide by plasma oxidization (Harrick Plasma; Ithaca, NY, USA). The assembled microfluidic devices were finally thermally aged at 65 °C overnight to limit uncured PDMS monomer leakage, and these devices were then ready to use.

2.4. Flow Control and Droplet Generation

For generating droplets on demand, a total of 19 pneumatic push-up valves on the microfluidic chip were programmatically controlled by an in-house written LabVIEW application which was interfaced to a custom manifold of solenoid switches (LHDA0533115H; the Lee Company, Westbrook, CT, USA) using a multifunction data acquisition system (PCI-6259, National Instruments). These solenoid valves were actuated by 5 V signals to controllably switch a pressurized nitrogen supply (25 psi), and only 13 solenoids were needed due to redundancy in operating some valves in the peristaltic pumps. For the periodic rinsing of the microdevice, the outlet could also be connected to a hand-held 100-mL syringe via Tygon tubing (0.02″ I.D. X 0.06″ O.D.; Cole-Parmer, Vernon Hills, IL, USA) to allow a small vacuum to be applied.

Droplets were generated with three-valve peristaltic pumps as described previously [7,9]. Oil segments were pumped in between each aqueous segment at a T-junction channel to form aqueous-in-oil droplets, and this formation was precisely controlled in an automated fashion using LabVIEW (Figure S1).

2.5. Programmable Merging of Droplets with Salt Water Electrodes

Droplets were merged with a 10 kHz alternating current (AC) signal of 500 V applied to nearby channels ("merging electrodes") filled with 5 M NaCl. The high voltage amplifier (Trek, Inc., Lockport, NY, USA; Model 2220) was controlled using an in-house written LabVIEW application. The merging region was widened when compared to incoming and outgoing channels to facilitate a slower migration and improved droplet contact for merging. This methodology was described in more detail by Sciambi and Abate [28].

User-defined time and channel programs for automatically building real-time five-point calibrations within sequentially merged droplets were preloaded into an in-house written LabVIEW application (Figure S2). Briefly, sequential groups of six droplets (2.77 nL each) were formed and separated in space to prevent group-to-group merging, and these six droplets were merged (16.6 nL in each larger droplet) using electrocoalescence downstream. As such, two types of oil segments were programmed: very short oil segments to keep droplets in the same group as close as possible, and longer oil segments to partition the sequential droplet groups. As discussed above, 84 possible solution combinations could be programmed into the finally merged droplet under the conditions investigated here. When applicable, the concentrations of an unknown sample could be determined in real time using continuous calibration curves, and signal drifts were corrected using the μChopper concept.

3. Results and Discussion

3.1. Microfluidic Device Design and Operation

As shown in Figure 1A, the microdevice was defined by several regions: (1) four different aqueous inlet reservoirs (colored) and an oil inlet reservoir (black); (2) T-junction channels for aqueous-in-oil droplet generation (colored and black); (3) pneumatic control channels (light gray) for automated chip operation through LabVIEW (National Instruments, Austin, TX, USA)), with some three-valve pumps integrated to improve efficiency; (4) salt water electrodes for droplet coalescence with a high AC voltage (two tones of dark blue); (5) a widened merging region near the salt water electrodes, at the sharpest electric field gradients; (6) a zig-zag channel (orange) for quickly and completely mixing reagents contained in droplets; and (7) a long incubation channel for storing and analyzing target droplets (orange). Regions (1) and (2) were valve-controlled AZ-defined rounded channels of ~40 μm depth, region (3) was SU-8 defined rectangular channels of ~20 μm depth, and regions (4)–(7) were SU-8 defined rectangular channels of ~55 μm depth.

In a typical assay workflow, calibration curves are regularly generated to allow the measurement and calculation of an unknown sample concentration. The conventional method is to quantify sequential standard solutions followed by each sample, then calculate the response curve and quantify samples post-measurement. Particularly when using expensive reagents—such as antibodies, protein standards, enzymes, bioconjugates, etc.—this traditional process not only wastes significant amounts of materials but also increases the workload of operators. To improve the accuracy and efficiency of building standard curves, we recently developed a six-channel μChopper to automatically carry out a continuous calibration mode [22], allowing the real-time determination of the slope, y-intercept, and correlation coefficients, along with unknown quantification. In this report, we improve upon this concept by introducing downstream droplet mergers via electrocoalescence, and we provide an even more precise control using on-chip pneumatic pumps.

To achieve full automation, we developed the device design in Figure 1A, which allows the programmable generation of droplets in various combinations from any of four input aqueous reservoirs. An example of a programmable calibration is shown in Figure 1B, where droplets containing dye solution (mimicking assay standards) and buffer are generated in various ratios. The images show droplets prior to merging into a single, larger droplet. Videos of programmable droplet formation (Video S1) and downstream merging (Video S2) are provided as supporting information. In this work, the total droplet number was limited due to the size of the coalescence region; however, this number could be increased by simply enlarging the dimensions of this region. With four input reservoirs and the total droplet count fixed at six, this system allowed for 84 possible solution combinations to be programmed into the finally merged droplet (16.6 nL). Notably, if the total droplet count were expanded to be one through six—easily accomplished with this device—there would be 209 possible solution combinations. The upper limit can be extended if the coalescing region is made larger; for example, if 24 total droplets were allowed (six from each reservoir), then 2400 combinations would be accessible.

3.2. Microdevice Characterization with Continuous Linear Calibration

To verify the automation capabilities of our device, fluorescein standard (165 nM), buffer, and an unknown fluorescein sample were loaded into reservoirs #1, #2, and #3, respectively (see Figure 1A). Five calibration standards were formulated by sequentially generating and coalescing groups of six droplets at varying ratios of fluorescein standard and buffer (1-5, 2-4, 3-3, 4-2, 5-1; akin to Figure 1B). For unknown measurements, a single larger-volume droplet was sampled from reservoir #3 and kept separate from the standard droplets. The blue and green traces in Figure 2A show a 20-min record of the raw fluorescence data measured at the incubation channel during the continuous calibration. To challenge the system during the continuous calibration, the excitation light was changed from a higher (initial settings) to a lower intensity (final settings). The unknown droplet's signal decreased by ~50% following this light intensity change. Since the signal from all of the calibration standards also decreased by the same proportion (~50%), the system allowed an accurate calibration to be maintained despite the challenge. Figure 2B shows a magnified view of the signal from one group of calibration standards and an unknown, and Figure 2C highlights the detector-dependent, low-frequency drift (noise) that can be corrected using the lock-in-based µChopper method.

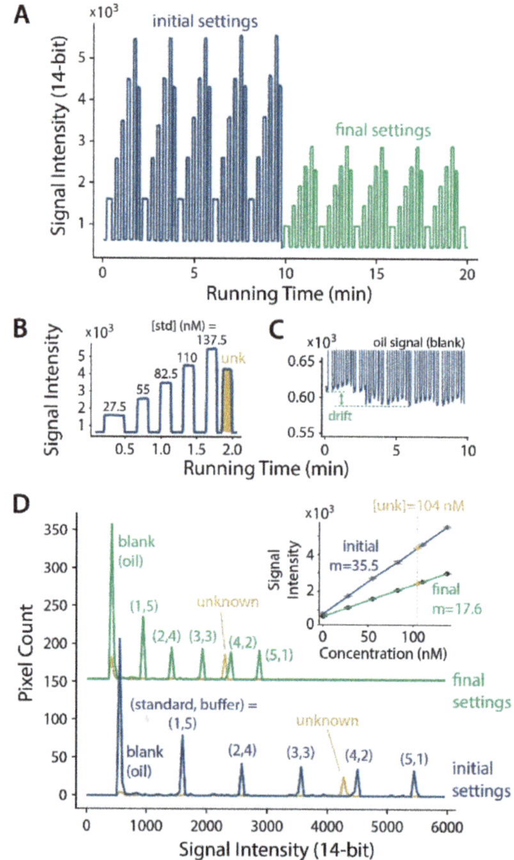

Figure 2. Continuous calibration with automated droplet formation and merging. (**A**) Raw fluorescence emission data shows that the droplet contents were programmable. Data is shown under initial settings at a higher excitation light intensity (blue) and with final settings after decreasing the light (green) in real time. (**B**) A magnified segment of this data, with pulses labeled using final, post-merge concentrations of fluorescein standard. Data from the unknown droplet is shaded in gold. (**C**) Magnified view of the oil signal shows a typical optical system drift that can be corrected using our µChopper method [8,12,22]. (**D**) Histogram analysis reveals the method's capability for a highly precise control of the droplet contents. The peaks are labeled with the pre-merge, programmed numbers of standard and buffer droplets. The inset shows the linear calibrations under the initial and final settings.

As shown in Figure 2D, the programmable droplet formation and merging enables a high-precision control over the final droplet composition. Essentially, this device operates as a microfluidic digital-to-analog converter, albeit at a relatively low resolution. The intensity histograms show that the fluorescence intensities were accurately controlled by the programmed ratios of the standard and buffer droplets, from (1,5) to (5,1). The initial settings (blue) showed that the unknown fluorescence was nearly as high as the (4,2) droplet, and after the light intensity challenge the final settings (green) showed the same order and position, just at a lower intensity. The inset calibration curves in Figure 2D show that the light intensity decrease mainly affected the calibration slope, while the unknown was determined to be at a concentration of 104 nM, independent of the excitation light intensity. These data point to a major advantage of continuous calibration, where the system can automatically adjust to drastic changes in the environmental conditions.

3.3. Unique Data Reshaping Using MATLAB Code

Since on-chip valves provide highly precise and programmable droplet intensities with repeatable timing, we surmised that the raw data could be readily reshaped into a more easily readable image format. Using a custom MATLAB (MathWorks, Natick, MA, USA) code (see Supporting Information), the raw data from Figure 2A was sliced into segments representing the repeating groups of larger droplets (116-second slices), and these slices were restacked over the running time of ~20 min and presented as the image in Figure 3A. The image intensity represents the 14-bit camera signal intensity using a custom colormap depicted in the legend. The five stripes of intense signal in this image represent a tracked intensity of each type of merged droplet, and the darker regions are the larger oil segments that separate them. For example, the blue stripe near the bottom of the image is a reshaping of data from multiple 27.5 nM standard droplets, where (standard,buffer) = (1,5), and this image allowed a facile tracking of their intensities over time. Indeed, the re-slicing of the image horizontally (shown above the image) gave the time traces of each standard (gold) and the unknown (blue). Again, with the change in light intensity, it is obvious that all calibration standards shifted along with the unknown to maintain calibration integrity. Conversely, by re-slicing the image vertically, the original data can be recovered, as shown at the right during both the initial (blue) and final (green) settings. This novel data reshaping approach is well-suited to an automated, droplet-based continuous calibration, and it was enabled by the precision of the valve-based control. It should be noted that a lock-in analysis was not yet applied to the data in this reshaped image, but further development of image analysis algorithms in the future will allow a lock-in analysis directly from these types of images.

This analysis gives a unique, visual means to showcase the system's ability to respond to environmental changes. Following the lock-in analysis, Figure 3B depicts the system's response to the challenge, where the major adjustment was a decrease in the slope (blue) of the linear calibration curve and a small change in the y-intercept (gold). The R^2 value (green) remained at a high level near 1.00 over the course of the sampling, and the unknown determination was steady at 104 nM despite the change in the excitation light intensity (blue data on the right, magnified to a 100–110 nM range).

3.4. Continuous Calibration Using a Nonlinear Homogeneous Immunoassay

Finally, we tested the performance of this droplet-based system using a more complex assay response. Using antibody-oligonucleotide conjugates (Ab-oligos) as probes (Figure 4B), where deoxyribonucleic acid (DNA) arms are labeled with a fluorophore and quencher, it is possible to quantify a protein analyte with a high specificity through a mix-and-read workflow [7,32,33] that is ideal for detection within droplets. However, the recovery of the sometimes small, unamplified signal changes can be challenging, particularly in the biologically relevant ranges for a hormone such as insulin (low ng mL^{-1}; pM to nM). We previously showed that our μChopper method provides a key enhancement to enable homogeneous immunoassays within droplets [8], and the combined techniques even allowed high-resolution sampling of insulin secretion from single pancreatic islets [7]. The disadvantages in these devices were that the mixing ratios of Ab-oligos and the sample were device-dependent, the assay timing was restricted by the flow rates of the on-chip pumps, and the calibrations had to be carried out before or after experiments in a serial fashion. Here (Figure 4A), we show that by forming aqueous-in-oil droplets of Ab-oligo probes (blue), insulin analyte (green), and buffer (gold) in a programmable way and then merging them downstream with integrated electrocoalescence, all of these aforementioned problems can be solved.

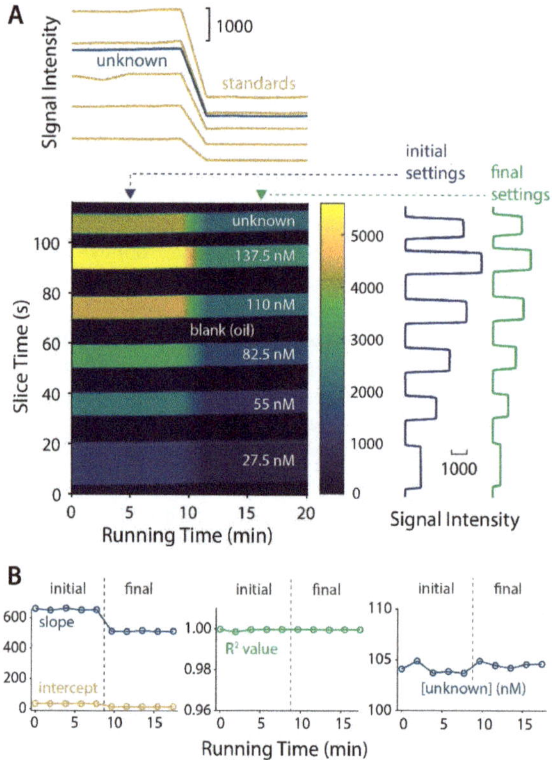

Figure 3. Data reshaping allowed for a unique visual inspection of the system, enabled by a precise droplet control with valves. (**A**) The raw data vector over time was reshaped into an image array using custom a MATLAB code, and image re-slicing permitted temporal tracking of each type of droplet (above) or original data recovery (right). (**B**) The system responded to the light intensity decrease by adjusting the calibration parameters, while the fit linearity and unknown determination were essentially unaffected.

By virtue of valve-based automation and downstream droplet merging, precious Ab-oligo reagents did not need to be diluted, premixed, or incubated with samples. For each measurement, a single drop of the stock Ab-oligo was sampled from inlet #1 (blue) and grouped with varying numbers of insulin (#3, green) and buffer (#4, gold) droplets (Figure 4A), and these six droplets were merged, mixed, and incubated downstream to allow for continuous calibration. The raw emission data in Figure 4C shows that the quenching within droplets was proportional to [insulin]. Also shown is the magnitude of the detector drift (two inset plots), which becomes highly significant compared to the signal changes in these homogeneous immunoassays at low analyte concentrations. In fact, the drift as high as ~640 intensity units was similar to the overall assay change for the full calibration range. Using the μChopper approach [8,12,22], these drifts were negated to give consistent calibration results over the entire experiment. The average curves are shown in Figure 4D, while the real-time curve parameters (linear fits versus $\log_{10}$[insulin]) are shown in Figure 4E. Continuous calibration allowed drifts and/or environmental changes to be negated, where the system continuously adjusted by modifying the slope and y-intercept. The concentration limit of detection (LOD$_{conc}$) was 2 ng mL^{-1} (300 pM), while the number of moles that were detectable (LOD$_{amt}$) was 5 amol (5 × 10^{-18} mol). This LOD is the best achieved to date for homogeneous insulin immunoassays using droplet-based microfluidics [7,8]. These data prove that the programmable device can give highly precise amounts of probe, calibration standards, and buffer—a significant improvement compared to the laminar flow sampling method in prior devices, where chip-to-chip variations were significant.

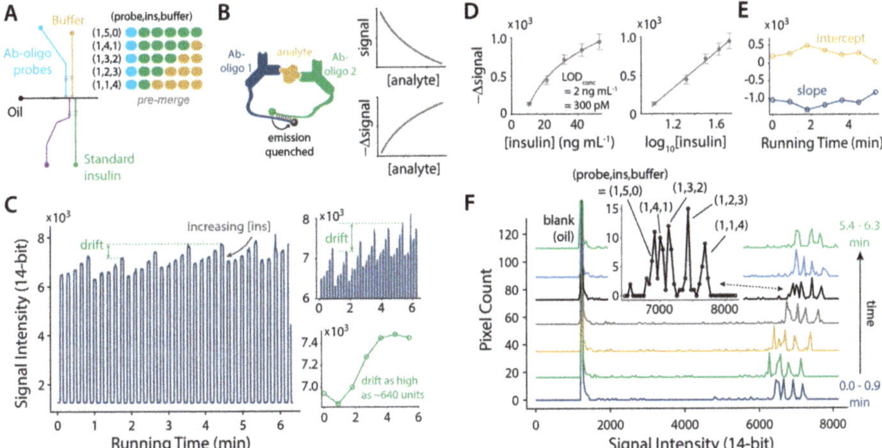

Figure 4. Automated homogeneous immunoassays in nanoliter droplets. (**A**) Device was operated with three inlets to program the pre-merge ratio of Ab-oligo probe, insulin, and buffer droplets. (**B**) Fluorescence-quenching-based homogeneous immunoassay with Ab-oligo probes. The signal quenching is proportional to the analyte concentration with a nonlinear response curve. (**C**) Raw emission data from the automated continuous calibration. The upper inset is a zoomed view of the detector drift, and the lower inset shows that the magnitude of the drift is similar to the overall assay response. (**D**) Lock-in detection with the μChopper method allows for a reliable correction and calibration. The signal change is shown versus [insulin] (left) and log$_{10}$[insulin] (right). LOD$_{conc}$ = 2 ng mL^{-1} = 300 pM, while LOD$_{amt}$ = 5 amol. (**E**) The continuous linear calibration parameters versus log$_{10}$[insulin] show the slope and y-intercept to be responsive to significant detector drifts. (**F**) The intensity histograms show that the assay responses over the 10–50 ng mL^{-1} insulin range were closely clustered, and drift could also be observed. The calibration standards followed the drift, giving reliable calibrations over time as in part (**D,E**).

The histograms shown in Figure 4F highlight some challenges that may arise using nonlinear, homogeneous immunoassays at low concentration ranges compared to simple direct fluorescence (as shown in Figure 2D). Calibration intensities were clustered together between 6000 and 8000 intensity units, and the drift could be readily observed. The inset magnified plot shows that the three highest [insulin] values—made with droplet ratios (1,5,0), (1,4,1), and (1,3,2)—were just barely resolved under these conditions. Fortunately, the μChopper method can compensate for these effects [8,12,22]. Overall, the data in Figure 4 show that automated sampling and downstream merging, when combined with lock-in detection, provide a highly reliable way to perform mix-and-read immunoassays in nanoliter droplets. The system can even be applied to the real-time quantification of proteins. These benefits were achieved with minimal user intervention, where the workflow consisted of adding merely three solutions to the inlet reservoirs before starting the system.

Lastly, an additional and noteworthy improvement is reduced cost. Using this device, the total volume used in one five-point calibration for the insulin immunoassay was 83.1 nL (five merged droplets of 16.6 nL). For the most expensive components, Ab-oligo probes, only 13.85 nL was required.

Because a five-point calibration in the standard 384-well plate version of the assay requires 2.25 µL of probes, our device reduced the needed volume 160-fold, which translates to an equivalent 160-fold reduction in cost for these precious bioconjugate components.

4. Conclusions

A fully automated microchip was introduced to precisely and rapidly form droplets of sequential calibration standards, allowing quantitative analyte measurements in the nanoliter range and in real time. The key novelty of this device was the integration of valve-based automation, on-chip droplet electrocoalescence, and µChopper data analysis. The user workflow was minimized to a few solution transfer steps at the beginning of the experiment, and cost reductions of more than two orders of magnitude (160-fold) were realized with homogeneous insulin immunoassays. Furthermore, full calibrations required <1 min, and this system also posted the lowest LOD achieved to date using droplet-based homogeneous immunoassays (insulin $LOD_{conc} = 2$ ng mL^{-1} = 300 pM; $LOD_{amt} = 5$ amol). The highly precise, programmable control also permitted unique data reshaping into images, with which lock-in detection or continuous referencing should be feasible through image analysis improvements.

Of course, some challenges remain with this system, which could be addressed in future work. To accommodate the serial sampling of multiple droplets and the formation of groups of calibration standards, the overall sample flow rate using on-chip pumps was lowered, causing about a one-order-of-magnitude loss in the temporal resolution of the sampling. Thus, the method timing is not yet competitive with our own state-of-the-art sampling resolution of 3.5 s [9]. While this issue is a cost of automation that is partially offset by the benefits, it could likely be improved by using gating valves [8] instead of full three-valve pumps. Finally, increases in the size or volume of the merging region would allow more droplets to be merged (>6). Changing this feature, along with increasing the input reservoir number, would exponentially increase the number of possible solution combinations, perhaps making it more palpable to refer to the chip as a microfluidic digital-to-analog converter.

Supplementary Materials: The following are available online at http://www.mdpi.com/2072-666X/11/6/620/s1, Figure S1: Automated µChopper device design and timing during operation, Figure S2: Programmatic flow chart of the LabVIEW application, Text: MATLAB code for reshaping data into an image, Video S1: "droplet generation", Video S2: "six merge".

Author Contributions: Conceptualization, C.J.E. and N.S.; methodology, C.J.E. and N.S.; software, C.J.E. and N.S.; validation, C.J.E. and N.S.; formal analysis, C.J.E. and N.S.; investigation, N.S.; resources, C.J.E.; data curation, N.S.; writing—original draft preparation, C.J.E. and N.S.; writing—review and editing, C.J.E. and N.S.; visualization, C.J.E. and N.S.; supervision, C.J.E.; project administration, C.J.E.; funding acquisition, C.J.E. All authors have read and agreed to the published version of the manuscript.

Funding: This research was funded by the National Institutes of Health (NIH), grant number R01 DK093810.

Conflicts of Interest: The authors declare no conflict of interest, and the funders had no role in the design of the study; in the collection, analyses, or interpretation of data; in the writing of the manuscript, or in the decision to publish the results.

References

1. Lorenz, R.M.; Fiorini, G.S.; Jeffries, G.D.; Lim, D.S.; He, M.; Chiu, D.T. Simultaneous generation of multiple aqueous droplets in a microfluidic device. *Anal. Chim. Acta* **2008**, *630*, 124–130. [CrossRef] [PubMed]
2. Lan, F.; Haliburton, J.R.; Yuan, A.; Abate, A.R. Droplet barcoding for massively parallel single-molecule deep sequencing. *Nat. Commun.* **2016**, *7*, 11784. [CrossRef] [PubMed]
3. Coelho, B.; Veigas, B.; Fortunato, E.; Martins, R.; Águas, H.; Igreja, R.; Baptista, P. Digital microfluidics for nucleic acid amplification. *Sensors* **2017**, *17*, 1495. [CrossRef] [PubMed]
4. Lee, J.Y.; Carugo, D.; Crake, C.; Owen, J.; de Saint Victor, M.; Seth, A.; Coussios, C.; Stride, E. Nanoparticle-loaded protein–polymer nanodroplets for improved stability and conversion efficiency in ultrasound imaging and drug delivery. *Adv. Mater.* **2015**, *27*, 5484–5492. [CrossRef]
5. Jin, S.H.; Lee, S.S.; Lee, B.; Jeong, S.G.; Peter, M.; Lee, C.S. Programmable static droplet array for the analysis of cell–cell communication in a confined microenvironment. *Anal. Chem.* **2017**, *89*, 9722–9729. [CrossRef]

6. Terekhov, S.S.; Smirnov, I.V.; Stepanova, A.V.; Bobik, T.V.; Mokrushina, Y.A.; Ponomarenko, N.A.; Belogurov, A.A.; Rubtsova, M.P.; Kartseva, O.V.; Gomzikova, M.O. Microfluidic droplet platform for ultrahigh-throughput single-cell screening of biodiversity. *Proc. Natl. Acad. Sci. USA* **2017**, *114*, 2550–2555. [CrossRef]

7. Li, X.; Hu, J.; Easley, C.J. Automated microfluidic droplet sampling with integrated, mix-and-read immunoassays to resolve endocrine tissue secretion dynamics. *Lab Chip* **2018**, *18*, 2926–2935. [CrossRef]

8. Negou, J.T.; Avila, L.A.; Li, X.; Hagos, T.M.; Easley, C.J. Automated microfluidic droplet-based sample chopper for detection of small fluorescence differences using lock-in analysis. *Anal. Chem.* **2017**, *89*, 6153–6159. [CrossRef]

9. Hu, J.; Li, X.; Judd, R.L.; Easley, C.J. Rapid lipolytic oscillations in ex vivo adipose tissue explants revealed through microfluidic droplet sampling at high temporal resolution. *Lab Chip* **2020**, *20*, 1503–1512. [CrossRef] [PubMed]

10. Easley, C.J.; Rocheleau, J.V.; Head, W.S.; Piston, D.W. Quantitative measurement of zinc secretion from pancreatic islets with high temporal resolution using droplet-based microfluidics. *Anal. Chem.* **2009**, *81*, 9086–9095. [CrossRef] [PubMed]

11. DeJournette, C.J.; Kim, J.; Medlen, H.; Li, X.; Vincent, L.J.; Easley, C.J. Creating biocompatible oil–water interfaces without synthesis: Direct interactions between primary amines and carboxylated perfluorocarbon surfactants. *Anal. Chem.* **2013**, *85*, 10556–10564. [CrossRef] [PubMed]

12. Deal, K.S.; Easley, C.J. Self-regulated, droplet-based sample chopper for microfluidic absorbance detection. *Anal. Chem.* **2012**, *84*, 1510–1516. [CrossRef] [PubMed]

13. Sciambi, A.; Abate, A.R. Accurate microfluidic sorting of droplets at 30 khz. *Lab Chip* **2015**, *15*, 47–51. [CrossRef] [PubMed]

14. Zhu, P.; Wang, L. Passive and active droplet generation with microfluidics: A review. *Lab Chip* **2017**, *17*, 34–75. [CrossRef] [PubMed]

15. Zeng, Y.; Shin, M.; Wang, T. Programmable active droplet generation enabled by integrated pneumatic micropumps. *Lab Chip* **2013**, *13*, 267–273. [CrossRef]

16. Chong, Z.Z.; Tan, S.H.; Gañán-Calvo, A.M.; Tor, S.B.; Loh, N.H.; Nguyen, N.-T. Active droplet generation in microfluidics. *Lab Chip* **2016**, *16*, 35–58. [CrossRef] [PubMed]

17. Doonan, S.R.; Lin, M.; Bailey, R.C. Droplet car-wash: Continuous picoliter-scale immunocapture and washing. *Lab Chip* **2019**, *19*, 1589–1598. [CrossRef] [PubMed]

18. O'Donovan, B.; Eastburn, D.J.; Abate, A.R. Electrode-free picoinjection of microfluidic drops. *Lab Chip* **2012**, *12*, 4029–4032. [CrossRef]

19. Li, X.P.; Brooks, J.C.; Hu, J.; Ford, K.I.; Easley, C.J. 3d-templated, fully automated microfluidic input/output multiplexer for endocrine tissue culture and secretion sampling. *Lab Chip* **2017**, *17*, 341–349. [CrossRef]

20. Chen, C.-T.; Lee, G.-B. Formation of microdroplets in liquids utilizing active pneumatic choppers on a microfluidic chip. *J. Microelectromech. Syst.* **2006**, *15*, 1492–1498. [CrossRef]

21. Godwin, L.A.; Deal, K.S.; Hoepfner, L.D.; Jackson, L.A.; Easley, C.J. Measurement of microchannel fluidic resistance with a standard voltage meter. *Anal. Chim. Acta* **2013**, *758*, 101–107. [CrossRef] [PubMed]

22. Negou, J.T.; Hu, J.; Li, X.; Easley, C.J. Advancement of analytical modes in a multichannel, microfluidic droplet-based sample chopper employing phase-locked detection. *Anal. Methods* **2018**, *10*, 3436–3443. [CrossRef] [PubMed]

23. Song, H.; Tice, J.D.; Ismagilov, R.F. A microfluidic system for controlling reaction networks in time. *Angew. Chem. Int. Ed.* **2003**, *42*, 768–772. [CrossRef] [PubMed]

24. Kang, D.-K.; Ali, M.M.; Zhang, K.; Huang, S.S.; Peterson, E.; Digman, M.A.; Gratton, E.; Zhao, W. Rapid detection of single bacteria in unprocessed blood using integrated comprehensive droplet digital detection. *Nat. Commun.* **2014**, *5*, 5427. [CrossRef]

25. Clark, A.M.; Sousa, K.M.; Jennings, C.; MacDougald, O.A.; Kennedy, R.T. Continuous-flow enzyme assay on a microfluidic chip for monitoring glycerol secretion from cultured adipocytes. *Anal. Chem.* **2009**, *81*, 2350–2356. [CrossRef]

26. Niu, X.; Gulati, S.; Edel, J.B.; Demello, A.J. Pillar-induced droplet merging in microfluidic circuits. *Lab Chip* **2008**, *8*, 1837–1841. [CrossRef]

27. Sander, J.S.; Erb, R.M.; Denier, C.; Studart, A.R. Magnetic transport, mixing and release of cargo with tailored nanoliter droplets. *Adv. Mater.* **2012**, *24*, 2582–2587. [CrossRef]

28. Sciambi, A.; Abate, A.R. Generating electric fields in pdms microfluidic devices with salt water electrodes. *Lab Chip* **2014**, *14*, 2605–2609. [CrossRef]

29. Sesen, M.; Alan, T.; Neild, A. Microfluidic on-demand droplet merging using surface acoustic waves. *Lab Chip* **2014**, *14*, 3325–3333. [CrossRef]

30. Doonan, S.R.; Bailey, R.C. K-channel: A multifunctional architecture for dynamically reconfigurable sample processing in droplet microfluidics. *Anal. Chem.* **2017**, *89*, 4091–4099. [CrossRef]

31. Erickstad, M.; Gutierrez, E.; Groisman, A. A low-cost low-maintenance ultraviolet lithography light source based on light-emitting diodes. *Lab Chip* **2015**, *15*, 57–61. [CrossRef] [PubMed]

32. Heyduk, E.; Dummit, B.; Chang, Y.-H.; Heyduk, T. Molecular pincers: Antibody-based homogeneous protein sensors. *Anal. Chem.* **2008**, *80*, 5152–5159. [CrossRef] [PubMed]

33. Hu, J.; Easley, C.J. Homogeneous assays of second messenger signaling and hormone secretion using thermofluorimetric methods that minimize calibration burden. *Anal. Chem.* **2017**, *89*, 8517–8523. [CrossRef] [PubMed]

Article

Microfluidic Droplet-Storage Array

Hoon Suk Rho [1,2] and Han Gardeniers [2,*]

[1] Department of Instructive Biomaterials Engineering, MERLN Institute for Technology-Inspired Regenerative Medicine, Maastricht University, 6229 ER Maastricht, The Netherlands; h.rho@maastrichtuniversity.nl

[2] Mesoscale Chemical Systems Group, MESA+ Institute, University of Twente, 7522 NB Enschede, The Netherlands

* Correspondence: j.g.e.gardeniers@utwente.nl; Tel.: +31-(0)53-489-4356

Received: 30 May 2020; Accepted: 22 June 2020; Published: 23 June 2020

Abstract: A microfluidic droplet-storage array that is capable of the continuous operation of droplet formation, storing, repositioning, retrieving, injecting and restoring is demonstrated. The microfluidic chip comprised four valve-assisted droplet generators and a 3 × 16 droplet-storage array. The integrated pneumatically actuated microvalves enable the precise control of aqueous phase dispensing, as well as carrier fluid flow path and direction for flexible manipulating water-in-oil droplets in the chip. The size of droplets formed by the valve-assisted droplet generators was validated under various operating conditions such as pressures for introducing solutions and dispensing time. In addition, flexible droplet addressing in the storage array was demonstrated by storing droplets with various numbers and compositions in different storage units as well as rearranging their stored positions. Moreover, serial injections of new droplets into a retrieved droplet from a storage unit was performed to show the potential of the platform in sequential dosing on incubated droplet-based reactors at the desired timeline. The droplet-storage array with great freedom and flexibility in droplet handling could be applied for performing complex chemical and biologic reactions, especially in which incubation and dosing steps are necessary.

Keywords: microfluidics; droplet array; microvalve

---

1. Introduction

Droplet-based microfluidic systems that manipulate nano- or picoliter droplets in microchannels have been highlighted in high-throughput chemical and biologic screening with rapid and robust reactions [1–4]. With the beneficial aspects of droplet-based reactors, e.g., extremely low sample consumption as well as flexibility in the size of the reactor, enormous droplet generators have been applied for various applications, including particle synthesis [5–7], cell analysis and nucleic acid amplification [8,9] and sequential biochemical reactions [10–12]. Besides this, advanced microfabrication and microengineering techniques have been exploited to design and manufacture innovative droplet devices for combinatorial screening [13–15], massive droplet production [16–18] and accurate droplet sorting [19–21].

One of the fundamental aspects of the droplet-based systems for analytical chemistry and biology is preparing libraries of samples in various compositions as well as concentrations. For varying the combination and concentration of reagents in a series of droplets, microfluidic mixers were integrated with droplet generators [5,10,11]. In the combined designs, several flowing streams were emerged into a microfluidic mixer and flowed into a carrier fluid flow to form droplets with various conditions. The other concept to create concentration gradients along droplets is merging droplets by synchronizing droplets in a microchannel network [22] or using hydrodynamics with microstructures designed to trap and release droplets [13,14,23,24].

Recently, highly integrated droplet-based microfluidic arrays have been developed for combinatorial screening in a single platform [13,14,25–30]. The microfluidic platforms combined broad functional spectrum in droplet-based microfluidics, e.g., concentration gradient generation and droplet formation and storage, for building high-throughput libraries in protein crystallization [25,26], single-cell analysis [13,14,28,29] and spheroids formation [27]. In particular, static droplet arrays [24,27–30] enabled controllable droplet generation and storage, as well as multiple chemical dose to stationary droplets in a relatively simple device geometry as a promising and inexpensive alternative to conventional high-throughput screening. Although the static droplet arrays showed great potential in performing complex combinatorial screening, the reactions are based on stationary droplets with predetermined volumes. Hence, the flexible droplet size-control and droplet retrieval for further on-chip and off-chip analysis remain challenging.

After the first introduction of Quake's valves, pneumatically actuated poly(dimethylsiloxane) (PDMS) monolithic valves fabricated by multilayer soft lithography [31], into droplet-based microfluidics [32], several multilayer PDMS devices have been presented to perform accurate handling of droplets for biochemical applications such as static fluorescence assay [33], nanoparticle synthesis [34] and single microbial cell screening [14]. The platforms showed the potential of the valve-assisted droplet generator to produce highly monodispersed droplets and combinatorial contents in a series of droplets. Furthermore, the accurate manipulation of complex fluid flows by multilayer devices, where tens- or hundreds of microvalves were integrated, in previous reports [35–37] showed promise to engineer an automated, multifunctional microfluidic droplet array.

Here, we discuss the development of a microfluidic droplet-storage array that enabled programmable droplet formation, multiple injections into generated droplets, addressing and incubating droplets in storage units, dosing additional droplets to stored droplets. The device comprised four valve-assisted droplet generators and an array of 48 droplet-storage units by integrating a forward–backward flow direction control and flow path changing with pneumatically actuated microfluidic valves. The flow rate ratio of a dispensing phase and a carrier fluid, as well as dispensing time, were characterized to generate water-in-oil (w/o) droplets with controlled volumes ranging from 13.2 ± 0.5 to 1204.8 ± 17.8 pL (n = 20). In addition, the repositioning of sets of stored droplets in the storage unit array was demonstrated to show the robust valve-assisted operation of flow direction and path control. Finally, the continuous operation of multiple droplet formation, storage, positioning, retrieval, and injecting was processed to present the great freedom and flexibility of the microfluidic chip in droplet handling. The microfluidic droplet-storage array may be applied for performing complex chemical and biologic reactions with tiny sample consumption, especially in which incubation and dosing steps are required, e.g., cell drug–dose response and multistep chemical synthesis.

2. Materials and Methods

2.1. Chip Design

The microfluidic droplet-storage array comprised 4 droplet generators and 48 droplet-storage units (Figure 1). The main carrier fluid flow channel was connected to two sets of an inlet and an outlet, and the carrier fluid flowed through the droplet generators and the storage unit array. Figure 1A shows the working process for droplet formation by using a microfluidic valve [32]. In each droplet generator, a pneumatically actuated microfluidic valve [31] was placed at the T-junction of an aqueous and the carrier fluid channels to control the connection between the two flows. The valve was normally closed to disconnect the two channels, and a volume of the aqueous phase was dispensed into the oil flow when the valve was open. Then the valve was closed again to create a segment of the aqueous phase in the carrier fluid flow. The size of the water-in-oil (w/o) droplet, the dispensed volume of the water phase, was determined by the flow rates of water and oil phases as well as the valve opening time.

Figure 1B presents the design of the inlet and outlet connection in the device for switching the flow direction of the carrier fluid (Figure 1B). On the two sides of the device, two sets of an inlet and an outlet were placed, and the channel connection was controlled by valves, which were designed at the junctions of channels. When the main oil flow channel was connected to the inlet near the droplet generators (cf. in #1) and the outlet after the incubation chambers (cf. out #1), the carrier fluid flowed from the droplet generators to incubation chambers (forward fluid flow). The oil flow direction was reversed to backward fluid flow by connecting the outlet before the droplet generators (cf. out #2) and the inlet after the chamber array (cf. in #2). The oil flow direction switching enabled the flexible addressing of droplets into the storage units and multiple droplet injections into a target droplet.

Each droplet-storage unit consisted of a bypass channel from the main carrier fluid flow channel, and two valves at the entrance and exit of the bypass channel, as illustrated in Figure 1C. By closing the valve near the bypass entrance and opening the other valve at the exit of the channel, droplets in the carrier fluid flowed into the bypass channel, the storage unit. Then, the droplets in the storage unit were trapped and isolated while the next droplets passed through the main oil channel by switching the valves on and off. The valves for isolating droplets in the 48 storage units were operated by microfluidic multiplexors [38]. The device contained 3 columns and 16 rows of the storage units within a chip dimension of 3.5 cm × 2.0 cm and 1–10 droplets, depending on the droplet size and droplet-to-droplet spacing, could be stored in each storage unit. Further integration of the storage units in a single chip would be possible by adding parallelized storage units and control channels for high-throughput analysis.

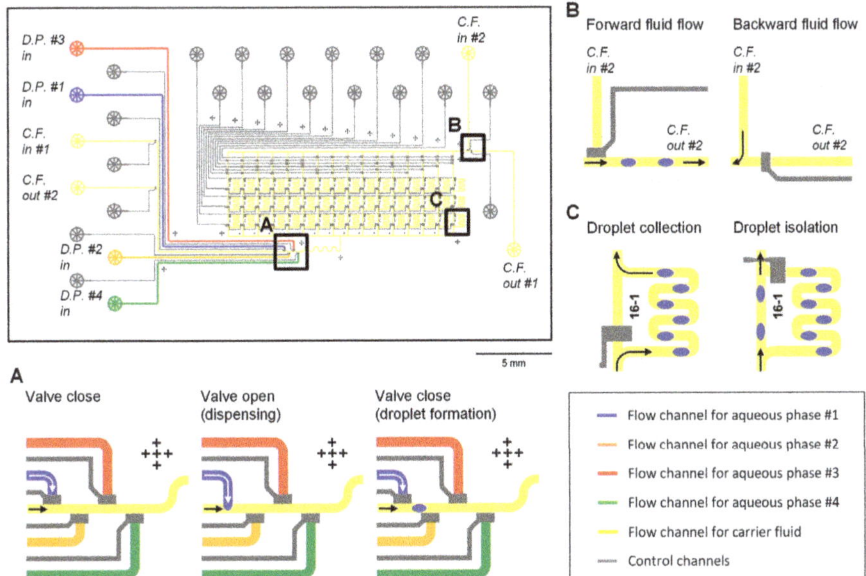

Figure 1. Design and operation of the microfluidic droplet-storage array. A computer-aided design shows the integration of 4 droplet generators and 48 droplet-storage units; (**A**) process flow of the droplet generation by using a microfluidic valve; (**B**) fluid flow direction control by switching a carrier fluid flow channel connection between the two sets of an inlet and an outlet; (**C**) valve operation to collect and isolate droplets in a droplet-storage unit.

2.2. Chip Fabrication

The microfluidic device was fabricated by multilayer soft lithography technique [31,39], and we followed a modified fabrication protocol based on our previous studies [36,37]. The PDMS device consisted of a top fluidic layer and a bottom control layer; the heights of fluid flow channels and control channels were 38 ± 2 μm and 18 ± 2 μm (n = 10), respectively.

2.3. Chip Operation

A pneumatic control system was used for operating fluid flow in the microfluidic droplet-storage array. Reagents were loaded into the flow channels by applying pressure from the backside of solutions [40], and microvalves were actuated by applying compressed nitrogen gas into the control ports. The pneumatic control system was automated by interconnecting 3/2-way solenoid valve manifolds, precision pressure regulators and an EasyPort USB digital I/O controller (all from Festo, Delft, The Netherlands) and controlled by a LabVIEW program (National Instruments Co., Austin, TX, USA). The valve operation for changing the fluid flow direction and droplet injection was controlled by sequencing pre-measured droplet moving time. However, a feedback control strategy may enable real-time sequencing by integrating a digital video processing software [41] with the operating setup.

2.4. Materials

Food dye solutions filtered with a 0.2-μm syringe filter (Whatman PLC, Sigma-Aldrich, Zwijndrecht, The Netherlands) and mineral oil containing 1.5% (*w/w*) Span 80 (all from Sigma-Aldrich, Zwijndrecht, The Netherlands) were used as the aqueous phases and the carrier fluid, respectively, for the generation of water-in-oil droplets. For sequential dilution by serial droplet injection, 1-g/L rhodamine B isothiocyanate-dextran (RITC-dextran, average molecule weight ~10,000, Sigma-Aldrich Chemie BV, Zwijndrecht, Netherlands) prepared in Milli-Q water (Millipore Co.) and Milli-Q water were used as a stock solution and a diluent, respectively.

2.5. Data Processing

A stereomicroscope (Motic SMZ171-TLED, LabAgency Benelux B.V., Dordrecht, The Netherlands) equipped with a CMOS camera (Moticam 3.0) and an inverted fluorescence microscope (Olympus IX73, Olympus Netherlands BV, Leiderdorp, The Netherlands) installed with a digital camera (ORCA-ER, Hamamatsu Photonics Deutschland GmbH, Herrsching, Germany) and an automatic XY-stage (99S000, Ludl Electronic Products, Ltd., NY, USA), were used for monitoring droplet generation and addressing in the device. An N 2.1 filter cube (excitation: BP 515–560 nm; emission: LP 590 nm) was used for observing RITC-dextran fluorescence signals with the fluorescence microscope. Acquired images and recorded movies were processed and analyzed by VirtualDub software (http://www.virtualdub.org/) and Image J software (http://rsb.info.nih.gov/ij/).

3. Results and Discussion

3.1. Droplet Generation with a Pneumatically Actuated Valve

The w/o droplets were formed at the T-junction of the aqueous phase and carrier fluid channels in the device by using a pneumatically actuated valve [32]. The sequence of the valve-assisted droplet formation in the droplet generator is shown in the time–series images in Figure 2A. Initially, pressures were applied for loading the aqueous phase and carrier fluid in the channels; however, the water flow was seized by closing the valve in the entrance of the aqueous phase channel. When the valve was open, the aqueous phase flowed into the main carrier fluid channel until the valve is closed again. Consequently, the dispensed volume of the aqueous phase created a water droplet in the oil flow. Hence, the size of the droplet was determined by the applied pressure for water and oil flows, as well as the opening time of the valve. The applied pressure ratio for water and oil flows (Pwater/Poil) and dispensing time for the droplet generation in Figure 2A are 1 and 167 ms, respectively.

For the calibration of operating conditions of the droplet generator, we created droplets under various dispensing times at constant applied pressure for water and oil phases, and different applied pressure ratios of water flow to oil flow while dispensing time was kept constant. Figure 2B shows droplet formation with various dispensing times ranging from 48 ms to 333 ms at a constant fluid flow condition ($P_{water}/P_{oil} = 1$). As an increased dispensing time, the formed droplet-size linearly increased. The relationship between the dispensing time and droplet volume at various fluid flow conditions is shown in Figure 2C (n = 20). The linear regressions with relatively small standard deviations represent the accurate droplet size-control and monodispersity of the formed droplets in the valve-assisted droplet generator. Supplementary Video S1 in Supplementary Materials shows the demonstration of valve-assisted droplet formation by varying dispensing time continuously.

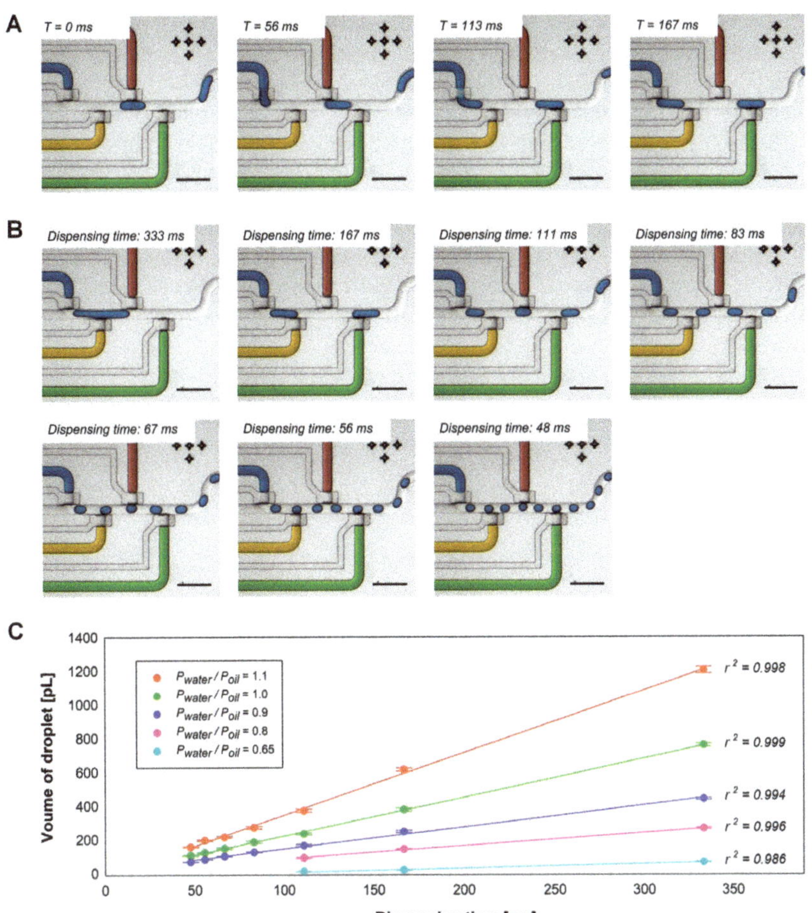

Figure 2. The w/o droplet generation with a pneumatically actuated microfluidic valve; (**A**) Step-by-step procedures of droplet formation at the constant droplet generation condition (P_{water}/P_{oil} = 1 and dispensing time = 167 ms); (**B**) Droplet size-control by varying the dispensing time at the constant fluid flow condition (P_{water}/P_{oil} = 1). Images captured from a recorded movie. Scale bars: 400 μm; (**C**) relationship between the dispensing time and the volume of formed droplets at various fluid flow conditions (n = 20).

3.2. Droplet Addressing by the Fluid Flow Direction Control

For demonstrating the droplet addressing capability of our microfluidic device, we relocated collected droplets in droplet-storage units (Figure 3). We generated and isolated nine sets of three water droplets formed with three different colored dye solutions, blue, red and yellow, into nine storage units. Initially, storage units in the first row, 1–1, 1–2 and 1–3, the second row, 2–1, 2–2 and 2–3, and the third rows, 3–1, 3–2 and 3–3, were filled with blue, red, yellow droplets, red, yellow and blue droplets and yellow, blue and red droplets, respectively (Figure 3A). Each storage unit contained one set of valves for bypass and collection of droplets into the unit, and the moving path of droplets was determined by the valve actuations. The flow direction of the carrier fluid was controlled by switching the connection of the two sets of an inlet and an outlet. For example, the droplets were moved to forward direction when the inlet of the carrier fluid near droplet generators and the outlet of the oil phase behind the chamber array were connected. On the contrast, the carrier fluid flowed backward with the connection of the oil inlet behind the chambers and the oil outlet closed to the droplet generators. For relocating droplets in storage units in the second row, blue droplets were retrieved from the storage unit 2–3 by connecting the bypass lines of 2–1 and 2–2 and the collection line of 2–3 with backward carrier fluid flow (Figure 3B). Then, the carrier fluid flow direction was changed to forward while the collection lines of 2–1, 2–2 and 2–3 were connected for collecting blue, red and yellow droplets in storage units 2–1, 2–2 and 2–3 (Figure 3C). By repeating the processes the yellow droplets in storage unit 3–1 were retrieved (Figure 3D) and pushed blue and red droplets in the storages 3–2 and 3–3 (Figure 3E) to make the same color order of stocked droplets, blue, red and yellow through the first, second and third columns (Figure 3F). The demonstration of relocating droplets in the storage units shows the capability of the device in collecting droplets in the array as well as freedom in setting the order of droplet-based reactors. The Supplementary Video S2 shows the procedure in real time.

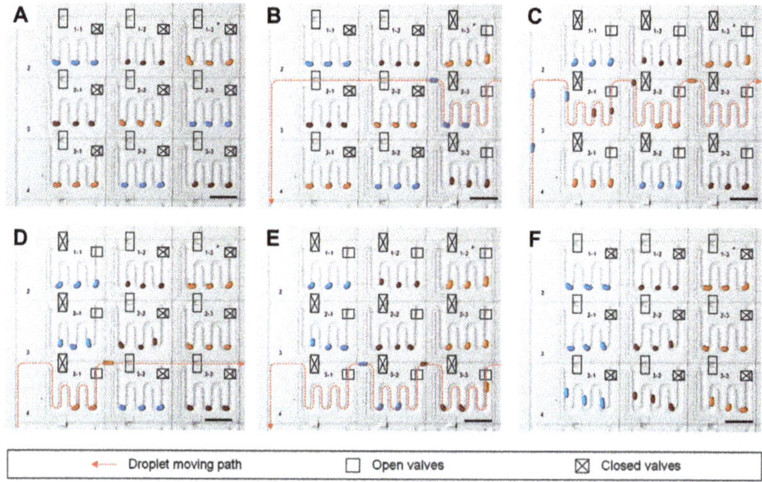

Figure 3. Droplet addressing in the storage unit array. (A) Nine sets of three droplets were isolated in a 3 × 3 storage unit array with random orders in droplet colors. (B–E) By moving the droplets with the control of carrier fluid flow direction as well as droplet moving path, (F) the color orders of droplets in each row rearranged to blue, red and yellow. Scale bars: 500 μm.

3.3. Serial Injection of Droplets into a Target Droplet

In the most chemical and biologic experiments, dilution and mixing of samples are fundamental operations for preparing, processing and analyzing reactions. Merging droplets is one of the most useful and practical operations in droplet-based reactors to vary the compositions and concentration of reagents for performing complex reactions [13,14,22–24,34]. Droplet-based microfluidic devices with continuous flows controlled the droplet merging by sequencing droplet formations [22] or integrating microstructures where droplets were trapped by hydrodynamics [13,14,23,24]. In valve-assisted droplet-based microfluidics, droplet merging was operated by synchronizing valve operation of in-line droplet generators [34].

The droplet-injection in our microfluidic droplet-storage array device is based on the synchronization of droplet generators; however, the forward–backward flow direction control on the carrier fluid enabled multiple injections of droplets into a formed droplet in a single droplet generator.

Figure 4A and Video S3 in Supplementary Materials show the operating procedure to perform serial droplet-injection into a target droplet. After the formation of a blue droplet, the droplet flowed back to the droplet generators by changing the flow direction of the carrier fluid to backward. Then, the oil flow direction was switched to forward direction again, and a red droplet was dispensed into the blue droplet. The same process was repeated for the sequential injection of red droplets into the blue droplet. Figure 4B shows the serial dilution of RITC-dextran by injecting Mill-Q water droplets into a preformed droplet containing 1-g/L RITC-dextran. The applied pressure ratio of water flow to oil flow and dispensing time were 1 and 111 ms, respectively. By repeating the Milli-Q droplet injection, the volume of the RITC-dextran droplet linearly increased with an increase of 236 ± 4 pL (Figure 4B(1)). Figure 4B(2) shows the relationship between the calculated concentration and measured the fluorescence intensity of the RITC-dextran droplet in the sequential dilution. The RITC-dextran fluorescence intensity in the droplet linearly decreased as droplet volume increased by the serial injection of Milli-Q water droplets.

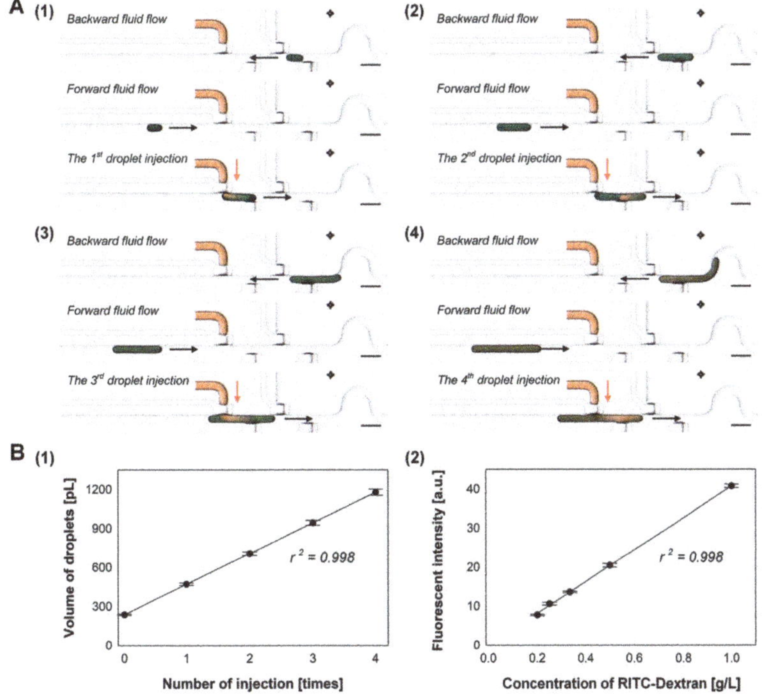

Figure 4. Serial injection of droplets into the target droplet in the droplet generator; (**A**) Process flow of the serial injection by changing the flow direction of the carrier fluid. Scale bars: 500 μm; (**B**) sequential dilution of RITC-dextran droplet (initial concentration of 1 g/L) by adding multiple Milli-Q water droplets (n = 3).

3.4. Continuous Processing of Droplet Formation, Addressing and Injection

To demonstrate the feasibility of the microfluidic device for creating a desired library of droplets in the storage unit array, we continuously processed multiple droplet generations with different numbers and reagents and droplet addressing in different storage units. In addition, we performed repositioning of droplets as well as injecting a new droplet into the droplet retrieved from the storage unit by controlling carrier fluid flow directions with valve operation.

Flexible droplet generation and storing in the microfluidic droplet-storage array is shown in Figure 5A. First, one red droplet and one green droplet were formed and collected in the storage units, 8–2 and 8–1 (Figure 5A(1,2)), then two red and two green droplets were generated and positioned in the storage units in the next row, 7–2 and 7–1 (Figure 5A(3)). The following droplets, one red and one green, were placed in the storage unit 6–1 together (Figure 5A(4)). Figure 5B shows the rearranging of the order of droplets by switching the oil flow path and flow. After generating two green droplets, a red droplet was formed and followed the green droplets. The green droplets were trapped in the storage unit 5–1, while the red droplet was flowed in the main oil channel next to the storage. Then we flowed one green droplet and inserted the red droplets between the two green droplets. Finally, the three droplets were isolated in the storage unit 5–1, in a new order, green–red–green. To demonstrate dosing a reagent into an incubated droplet, we added a blue droplet into the stored green droplet (Figure 5C). The forward–backward oil flow direction control enabled the operations of retrieving the green droplet from the storage unit (Figure 5C(1)), placing to droplet generator (Figure 5C(2)), adding a new droplet (Figure 5C(3)) and restoring into the storage unit (Figure 5C(4)). The Video S4 in Supplementary Materials shows the real-time operation of the processes.

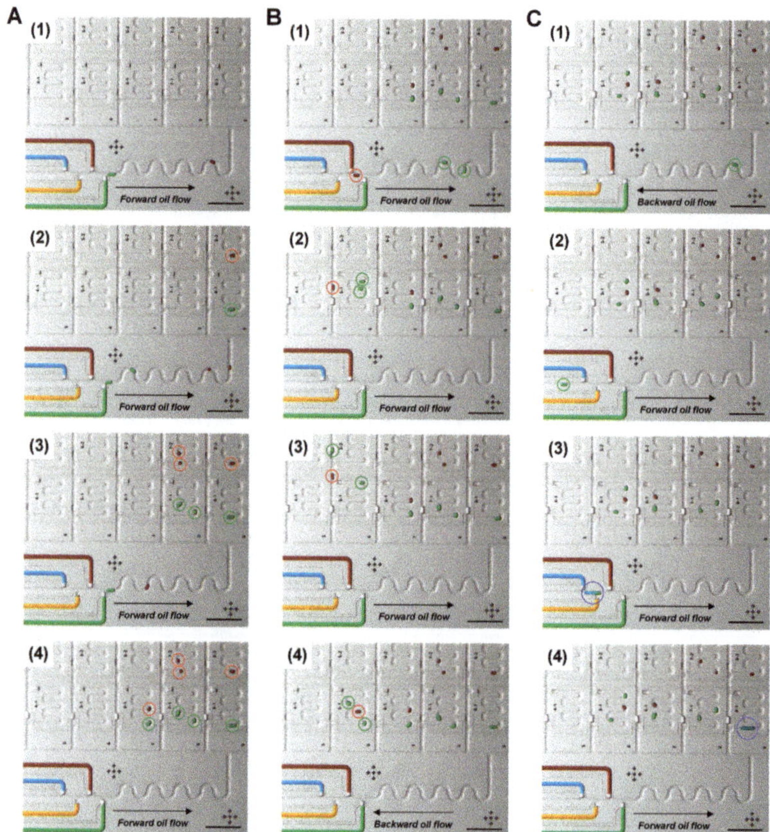

Figure 5. Continuous processes of (**A**) droplet formation and storing, (**B**) repositioning and (**C**) injection and restoring. Scale bars: 1 mm.

The other attractive function of the device for long-term incubation of droplets in the storage array is shaking droplets. The pristine PDMS shows a hydrophobic characteristic with a water contact angle of 100–112° [42]; however, the wetting of the aqueous phase onto the PDMS surface is still challenging for PDMS-based w/o droplet generators [43]. Although PDMS surface treatment with a commercial water repellent Aquapel [43,44] may reduce the water wetting on the PDMS surface, this non-permanent treatment is limited for the long-term storage of water droplets in PDMS channel without the loss of the water droplet volume. Our microfluidic droplet-storage device is capable of switching the main oil flow direction continuously by automated valve operation. Hence, water droplets collected in storage units can be shaken to prevent droplet settling in contact with the PDMS surface that may result in reducing the droplet volume during long-term incubation (Video S5 in Supplementary Materials). In non-treated PDMS channels, the volume loss of w/o droplets was 25% ± 3% of the initial volume of droplets with shaking while the droplet volume reduced 73% ± 2% without shaking after 5 h incubation at room temperature (Figure S1 in Supplementary Materials). Furthermore, the function of droplet shaking may be useful for conducting biochemical reactions, where agitation plays a critical role, such as protein fibrillation and aggregation [12,45,46].

4. Conclusions

In this work, we established a microfluidic droplet incubation chamber array to combine programmable droplet formation, multiple injections into formed droplets, addressing droplets into incubation chambers, dosing additional droplets to incubated droplets in a single device. All the droplet handling processes were performed by automated microfluidic valve control. Droplet formation with an accurate size control was validated under various valve operating conditions and applied pressure for loading solutions. In addition, flexible droplet addressing and multiple dosing on a formed droplet were demonstrated by multiplexing integrated microvalves and the forward–backward flow direction control. The microfluidic droplet incubation chamber array may be an attractive tool for performing complex chemical and biologic reactions with extremely small sample consumption, particularly in which incubation and dosing steps are required.

Supplementary Materials: The following are available online at http://www.mdpi.com/2072-666X/11/6/608/s1, Video S1: Valve-assisted droplet formation, Video S2: Droplet addressing in a storage array, Video S3: Serial injection of droplets, Video S4: Journey of droplets on a chip, Video S5: Shaking droplets in a storage unit and Figure S1: Volume loss of w/o droplets in a non-treated PDMS channel.

Author Contributions: Conceptualization, H.S.R. and H.G.; methodology, H.S.R. and H.G.; software, H.S.R.; validation, H.S.R.; formal analysis, H.S.R.; investigation, H.S.R. and H.G.; writing—original draft preparation, All authors have read and agreed to the published version of the manuscript.

Funding: This research was funded by the BE-BASIC FOUNDATION project number FS2.003 (funded by THE MINISTRY OF ECONOMIC AFFAIRS OF THE NETHERLANDS, Grant Number FES0905).

Acknowledgments: H.S.R. gratefully acknowledges the Innovational Research Incentives Scheme Vidi (# 15,604) of the Netherlands Organization for Scientific Research (NWO).

Conflicts of Interest: The authors declare no conflicts of interest.

References

1. DeMello, A.J. Control and detection of chemical reactions in microfluidic systems. *Nature* **2006**, *442*, 394–402. [CrossRef] [PubMed]
2. Teh, S.Y.; Lin, R.; Hung, L.H.; Lee, A.P. Droplet microfluidics. *Lab Chip* **2008**, *8*, 198–220. [CrossRef] [PubMed]
3. Huebner, A.; Sharma, S.; Srisa-Art, M.; Hollfelder, F.; Edel, J.B.; DeMello, A.J. Microdroplets: A sea of applications? *Lab Chip* **2008**, *8*, 1244–1254. [CrossRef]
4. Pompano, R.R.; Liu, W.; Du, W.; Ismagilov, R.F. Microfluidics using spatially defined arrays of droplets in one, two, and three dimensions. *Annu. Rev. Anal. Chem.* **2011**, *4*, 59–81. [CrossRef] [PubMed]
5. Shestopalov, I.; Tice, J.D.; Ismagilov, R.F. Multi-step synthesis of nanoparticles performed on millisecond time scale in a microfluidic droplet-based system. *Lab Chip* **2004**, *4*, 316–321. [CrossRef] [PubMed]
6. Millman, J.R.; Bhatt, K.H.; Prevo, B.G.; Velev, O.D. Anisotropic particle synthesis in dielectrophoretically controlled microdroplet reactors. *Nat. Mater.* **2005**, *4*, 98–102. [CrossRef] [PubMed]
7. Chan, E.M.; Alivisatos, A.P.; Mathies, R.A. High-temperature microfluidic synthesis of CdSe nanocrystals in nanoliter droplets. *J. Am. Chem. Soc.* **2005**, *127*, 13854–13861. [CrossRef] [PubMed]

8.   Brouzes, E.; Medkova, M.; Savenelli, N.; Marran, D.; Twardowski, M.; Hutchison, J.B.; Rothberg, J.M.; Link, D.R.; Perrimon, N.; Samuels, M.L. Droplet microfluidic technology for single-cell high-throughput screening. *Proc. Natl. Acad. Sci. USA* **2009**, *106*, 14195–14200. [CrossRef]

9.   Agresti, J.J.; Antipov, E.; Abate, A.R.; Ahn, K.; Rowat, A.C.; Baret, J.C.; Marquez, M.; Klibanov, A.M.; Griffiths, A.D.; Weitz, D.A. Ultrahigh-throughput screening in drop-based microfluidics for directed evolution. *Proc. Natl. Acad. Sci. USA* **2010**, *107*, 4004–4009. [CrossRef]

10.   Bringer, M.R.; Gerdts, C.J.; Song, H.; Tice, J.D.; Ismagilov, R.F. Microfluidic systems for chemical kinetics that rely on chaotic mixing in droplets. *Philos. Trans. R. Soc. A Math. Phys. Eng. Sci.* **2004**, *362*, 1087–1104. [CrossRef] [PubMed]

11.   Bui, M.P.N.; Li, C.A.; Han, K.N.; Choo, J.; Lee, E.K.; Seong, G.H. Enzyme kinetic measurements using a droplet-based microfluidic system with a concentration gradient. *Anal. Chem.* **2011**, *83*, 385–390. [CrossRef] [PubMed]

12.   Knowles, T.P.J.; White, D.A.; Abate, A.R.; Agresti, J.J.; Cohen, S.I.A.; Sperling, R.A.; De Genst, E.J.; Dobson, C.M.; Weitz, D.A. Observation of spatial propagation of amyloid assembly from single nuclei. *Proc. Natl. Acad. Sci. USA* **2011**, *108*, 14746–14751. [CrossRef] [PubMed]

13.   Padmanabhan, S.; Misteli, T.; Devoe, D.L. Controlled droplet discretization and manipulation using membrane displacement traps. *Lab Chip* **2017**, *17*, 3717–3724. [CrossRef]

14.   Jang, S.; Lee, B.; Jeong, H.H.; Jin, S.H.; Jang, S.; Kim, S.G.; Jung, G.Y.; Lee, C.S. On-chip analysis, indexing and screening for chemical producing bacteria in a microfluidic static droplet array. *Lab Chip* **2016**, *16*, 1909–1916. [CrossRef] [PubMed]

15.   Cole, R.H.; Tang, S.Y.; Siltanen, C.A.; Shahi, P.; Zhang, J.Q.; Poust, S.; Gartner, Z.J.; Abate, A.R. Printed droplet microfluidics for on demand dispensing of picoliter droplets and cells. *Proc. Natl. Acad. Sci. USA* **2017**, *114*, 8728–8733. [CrossRef] [PubMed]

16.   Amstad, E.; Chemama, M.; Eggersdorfer, M.; Arriaga, L.R.; Brenner, M.P.; Weitz, D.A. Robust scalable high throughput production of monodisperse drops. *Lab Chip* **2016**, *16*, 4163–4172. [CrossRef]

17.   Yadavali, S.; Jeong, H.H.; Lee, D.; Issadore, D. Silicon and glass very large scale microfluidic droplet integration for terascale generation of polymer microparticles. *Nat. Commun.* **2018**, *9*, 1–9. [CrossRef]

18.   Tang, S.Y.; Wang, K.; Fan, K.; Feng, Z.; Zhang, Y.; Zhao, Q.; Yun, G.; Yuan, D.; Jiang, L.; Li, M.; et al. High-Throughput, Off-Chip Microdroplet Generator Enabled by a Spinning Conical Frustum. *Anal. Chem.* **2019**, *91*, 3725–3732. [CrossRef]

19.   Baret, J.C.; Miller, O.J.; Taly, V.; Ryckelynck, M.; El-Harrak, A.; Frenz, L.; Rick, C.; Samuels, M.L.; Hutchison, J.B.; Agresti, J.J.; et al. Fluorescence-activated droplet sorting (FADS): Efficient microfluidic cell sorting based on enzymatic activity. *Lab Chip* **2009**, *9*, 1850–1858. [CrossRef]

20.   Caen, O.; Schütz, S.; Jammalamadaka, M.S.S.; Vrignon, J.; Nizard, P.; Schneider, T.M.; Baret, J.C.; Taly, V. High-throughput multiplexed fluorescence-activated droplet sorting. *Microsyst. Nanoeng.* **2018**, *4*, 1–10. [CrossRef]

21.   Isozaki, A.; Nakagawa, Y.; Loo, M.H.; Shibata, Y.; Tanaka, N.; Setyaningrum, D.L.; Park, J.-W.; Shirasaki, Y.; Mikami, H.; Huang, D.; et al. Sequentially addressable dielectrophoretic array for high-throughput sorting of large-volume biological compartments. *Sci. Adv.* **2020**, *6*, eaba6712. [CrossRef] [PubMed]

22.   Hong, J.; Choi, M.; Edel, J.B.; Demello, A.J. Passive self-synchronized two-droplet generation. *Lab Chip* **2010**, *10*, 2702–2709. [CrossRef] [PubMed]

23.   Niu, X.; Gulati, S.; Edel, J.B.; Demello, A.J. Pillar-induced droplet merging in microfluidic circuits. *Lab Chip* **2008**, *8*, 1837–1841. [CrossRef] [PubMed]

24.   Bithi, S.S.; Vanapalli, S.A. Behavior of a train of droplets in a fluidic network with hydrodynamic traps. *Biomicrofluidics* **2010**, *4*, 044110. [CrossRef]

25.   Lau, B.T.C.; Baitz, C.A.; Dong, X.P.; Hansen, C.L. A complete microfluidic screening platform for rational protein crystallization. *J. Am. Chem. Soc.* **2007**, *129*, 454–455. [CrossRef]

26.   Yang, C.G.; Liu, Y.H.; Di, Y.Q.; Xu, Z.R. Generation of two-dimensional concentration-gradient droplet arrays on a two-layer chip for screening of protein crystallization conditions. *Microfluid. Nanofluidics* **2015**, *18*, 493–501. [CrossRef]

27.   McMillan, K.S.; Boyd, M.; Zagnoni, M. Transitioning from multi-phase to single-phase microfluidics for long-term culture and treatment of multicellular spheroids. *Lab Chip* **2016**, *16*, 3548–3557. [CrossRef]

28. Bithi, S.S.; Vanapalli, S.A. Microfluidic cell isolation technology for drug testing of single tumor cells and their clusters. *Sci. Rep.* **2017**, *7*, 41707. [CrossRef]

29. Hassanzadeh-Barforoushi, A.; Law, A.M.K.; Hejri, A.; Asadnia, M.; Ormandy, C.J.; Gallego-Ortega, D.; Ebrahimi Warkiani, M. Static droplet array for culturing single live adherent cells in an isolated chemical microenvironment. *Lab Chip* **2018**, *18*, 2156–2166. [CrossRef]

30. Aubry, G.; Lu, H. Droplet array for screening acute behaviour response to chemicals in: Caenorhabditis elegans. *Lab Chip* **2017**, *17*, 4303–4311. [CrossRef]

31. Unger, M.A.; Chou, H.P.; Thorsen, T.; Scherer, A.; Quake, S.R. Monolithic microfabricated valves and pumps by multilayer soft lithography. *Science* **2000**, *288*, 113–116. [CrossRef]

32. Zeng, S.; Li, B.; Su, X.; Qin, J.; Lin, B. Microvalve-actuated precise control of individual droplets in microfluidic devices. *Lab Chip* **2009**, *9*, 1340–1343. [CrossRef] [PubMed]

33. Tangen, U.; Sharma, A.; Wagler, P.; McCaskill, J.S. On demand nanoliter-scale microfluidic droplet generation, injection, and mixing using a passive microfluidic device. *Biomicrofluidics* **2015**, *9*, 014119. [CrossRef] [PubMed]

34. Dai, J.; Yang, X.; Hamon, M.; Kong, L. Particle size controlled synthesis of CdS nanoparticles on a microfluidic chip. *Chem. Eng. J.* **2015**, *280*, 385–390. [CrossRef]

35. Melin, J.; Quake, S.R. Microfluidic large-scale integration: The evolution of design rules for biological automation. *Annu. Rev. Biophys. Biomol. Struct.* **2007**, *36*, 213–231. [CrossRef]

36. Rho, H.S.; Yang, Y.; Hanke, A.T.; Ottens, M.; Terstappen, L.W.M.M.; Gardeniers, H. Programmable v-type valve for cell and particle manipulation in microfluidic devices. *Lab Chip* **2016**, *16*, 305–311. [CrossRef]

37. Rho, H.S.; Veltkamp, H.W.; Hanke, A.T.; Ottens, M.; Breukers, C.; Habibović, P.; Gardeniers, H. Systematic investigation of insulin fibrillation on a chip. *Molecules* **2020**, *25*, 1380. [CrossRef]

38. Thorsen, T.; Maerkl, S.J.; Quake, S.R. Microfluidic large-scale integration. *Science* **2002**, *298*, 580–584. [CrossRef]

39. Xia, Y.; Whitesides, G.M. Soft lithography. *Annu. Rev. Mater. Sci.* **1998**, *28*, 153–184. [CrossRef]

40. Rho, H.S.; Yang, Y.; Veltkamp, H.; Gardeniers, H. Direct Delivery of Reagents from a Pipette Tip to a PDMS Microfluidic Device. Available online: https://blogs.rsc.org/chipsandtips/2015/10/09/direct-delivery-of-reagents-from-a-pipette-tip-to-a-pdms-microfluidic-device/ (accessed on 9 October 2015).

41. Basu, A.S. Droplet morphometry and velocimetry (DMV): A video processing software for time-resolved, label-free tracking of droplet parameters. *Lab Chip* **2013**, *13*, 1892–1901. [CrossRef]

42. Bodas, D.; Khan-Malek, C. Formation of more stable hydrophilic surfaces of PDMS by plasma and chemical treatments. *Microelectron. Eng.* **2006**, *83*, 1277–1279. [CrossRef]

43. Mazutis, L.; Gilbert, J.; Ung, W.L.; Weitz, D.A.; Griffiths, A.D.; Heyman, J.A. Single-cell analysis and sorting using droplet-based microfluidics. *Nat. Protoc.* **2013**, *8*, 870–891. [CrossRef] [PubMed]

44. Sabhachandani, P.; Motwani, V.; Cohen, N.; Sarkar, S.; Torchilin, V.; Konry, T. Generation and functional assessment of 3D multicellular spheroids in droplet based microfluidics platform. *Lab Chip* **2016**, *16*, 497–505. [CrossRef] [PubMed]

45. Brange, J.; Andersen, L.; Laursen, E.D.; Meyn, G.; Rasmussen, E. Toward understanding insulin fibrillation. *J. Pharm. Sci.* **1997**, *86*, 517–525. [CrossRef] [PubMed]

46. Fink, A.L. The aggregation and fibrillation of α-synuclein. *Acc. Chem. Res.* **2006**, *39*, 628–634. [CrossRef] [PubMed]

 *micromachines*

Article

Generation of Ultra-Thin-Shell Microcapsules Using Osmolarity-Controlled Swelling Method

Jianhua Guo, Lihua Hou, Junpeng Hou, Jiali Yu and Qingming Hu *

School of Mechatronics Engineering, Qiqihar University, Wenhua Street 42, Qiqihar 161006, Heilongjiang, China; guo1034@sina.com (J.G.); hlh0688@126.com (L.H.); houjunpengyjs@163.com (J.H.); futuredreams520@sina.com (J.Y.)

* Correspondence: mehuqm@qqhru.edu.cn; Tel.: +86-452-2738-241

Received: 14 March 2020; Accepted: 21 April 2020; Published: 23 April 2020

Abstract: Microcapsules are attractive core-shell configurations for studies of controlled release, biomolecular sensing, artificial microbial environments, and spherical film buckling. However, the production of microcapsules with ultra-thin shells remains a challenge. Here we develop a simple and practical osmolarity-controlled swelling method for the mass production of monodisperse microcapsules with ultra-thin shells via water-in-oil-in-water (W/O/W) double-emulsion drops templating. The size and shell thickness of the double-emulsion drops are precisely tuned by changing the osmotic pressure between the inner cores and the suspending medium, indicating the practicability and effectiveness of this swelling method in tuning the shell thickness of double-emulsion drops and the resultant microcapsules. This method enables the production of microcapsules even with an ultra-thin shell less than hundreds of nanometers, which overcomes the difficulty in producing ultra-thin-shell microcapsules using the classic microfluidic emulsion technologies. In addition, the ultra-thin-shell microcapsules can maintain their intact spherical shape for up to 1 year without rupturing in our long-term observation. We believe that the osmolarity-controlled swelling method will be useful in generating ultra-thin-shell polydimethylsiloxane (PDMS) microcapsules for long-term encapsulation, and for thin film folding, buckling and rupturing investigation.

Keywords: microcapsules; double-emulsion drops; osmotic pressure; ultra-thin-shell; microfluidics

---

1. Introduction

A microcapsule is a micrometer-scale core-shell structure, with compartments encapsulated in a solid shell. This kind of core-shell configuration not only can protect the core materials from external disturbance or even contamination, but can also perform the on-demand delivery and release of the aqueous core under various external stimuli [1,2]. Thus, microcapsules can serve as model systems in various applications, such as drug delivery and controlled release [3–7], photonic capsule sensors [8–10], artificial microbial environments [11,12], foods [13,14], and biomolecular sensing [15,16].

Microfluidics technology supplies a simple and effective method for the production of various microcapsules with well-defined compositions and structures via double-emulsion drops templating [1,17]. For instance, by solidifying the liquid shell of the double-emulsion drops using temperature, pH, chemistry or light-sensitive materials, microcapsules with controlled release behaviors can be mass produced [18–25]. A porous shell endows microcapsules with a size-selective permeability [26]. Multicompartment microcapsules with multi-cores or capsule-in-capsule structures are designed for co-encapsulation and diverse programmable sequential release [27–31]. Smart microcapsules, which have magnetic materials embedded on the shell or in the cores, have been used for direction-specific delivery and release [32–34].

In particular, microcapsules with ultra-thin shells are critical templates for the above mentioned applications, and for studies of folding and buckling behaviors of spherical thin films [35]. However, the production of microcapsules with ultra-thin shells remains a challenge. It is very difficult to fabricate a double-emulsion drop with an ultra-thin shell less than several hundred nanometers using classic capillary microfluidic devices. There have been some configurations of capillary microfluidics to fabricate such ultra-thin-shell double-emulsion drops and microcapsules [36–42]. For example, an ultra-thin middle layer of double-emulsion drops can be stably created by squeezing a thin layer of middle fluid between the inner wall of the capillary and the innermost fluid [36]. The middle fluid needs to have high affinity to the capillary wall. The phase compositions and wettability of the middle fluid, however, are strictly required to create the thin shell. Therefore, developing an alternative method with simple configuration and practical performance for mass production of ultra-thin-shell microcapsules is of great significance.

In this study, we demonstrate an easy and practical method to generate highly monodisperse ultra-thin-shell microcapsules via osmolarity-controlled swelling behaviors of water-in-oil-in-water (W/O/W) double-emulsion drops. To achieve this, a classic capillary microfluidic device is firstly designed to fabricate monodisperse W/O/W double-emulsion drops, as shown in Figure 1a. Then, an osmolarity-controlled swelling method is developed to make the shell thinner by increasing the drop diameter, as the schematic illustration shown in Figure 1b. Here, the swelling process of the emulsion drops is simply controlled by the osmotic pressure between the inner aqueous and the suspending medium. The shell material, polydimethylsiloxane (PDMS), enables the transportation of water into or out of the drops when they are subjected to an osmotic pressure difference [43,44]. The shell thickness of the double-emulsion drop after swelling is figured out by measuring the volume of the PDMS shell phase after an electro-triggered rupturing, the value of which can be several hundreds of nanometers or less. Finally, the double-emulsion drops are transformed into ultra-thin-shell microcapsules by a thermal curing method to solidify the PDMS shell phase. This method enables the mass production of microcapsules with ultra-thin shells, and the shell thickness can be controlled precisely.

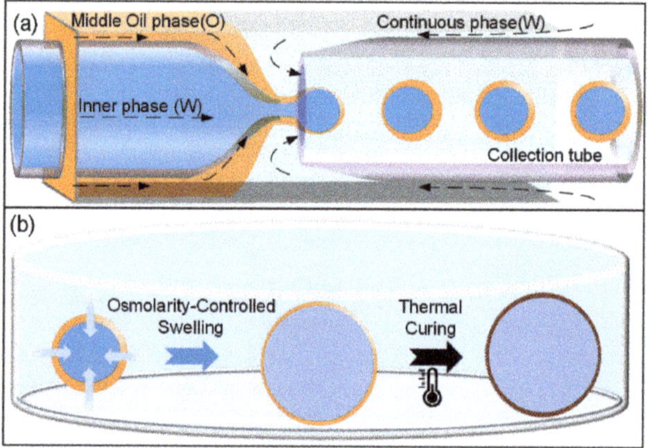

Figure 1. (a) Schematic illustration of the capillary microfluidic device for generating water-in-oil-in-water (W/O/W) double-emulsion drops; (b) schematic illustrations showing the osmolarity-controlled swelling process of W/O/W double-emulsion drop, and the thermal curing process from liquid emulsion drop to solid microcapsule.

2. Materials and Methods

2.1. Materials

Polyvinyl alcohol (PVA, 87–89% hydrolyzed, average Mw = 13,000–23,000), potassium chloride (KCl), octadecyltrichlorosilane (OTS), and methylene blue were purchased from Sigma-Aldrich (St. Louis, MO, USA). PDMS kit (Sylgard 184) and silicone oil (50 cSt, PMX-200) were purchased from

Dow Corning (Midland, MI, USA). As the outer continuous phase (W_{outer}), aqueous solution of 5 wt% PVA was used. PVA served as a surfactant in this case. As the middle oil phase (O_{middle}), a mixture of PDMS and silicone oil with the volume ratio of 3:1 was employed, in which 10 wt% curing agent was added for solidifying the double-emulsion drops into microcapsules. As the inner phase (W_{inner}), aqueous solution of 1 wt% PVA was used. Depending on different experiments, KCl was added into the inner phase and suspending medium to adjust the osmotic pressure between the inner and outer of the PDMS shell. In some experiments, methylene blue was dissolved into the inner core as a dye for better visualization. All water used in this study was deionized (DI) water unless otherwise noted.

2.2. Fabrication of the Glass Capillary Microfluidic Device

A classic glass-capillary microfluidic device was used to fabricate the W/O/W double-emulsion drops, as described previously [17,45]. Briefly, two circular capillary tubes (ID of 0.58 mm, OD of 1.03 mm, World Pricision Instrument Inc., Sarasota, FL, USA) were given tapered openings of 38 and 170 μm in diameter using a micropipette puller (P-97, Sutter Instrument Inc., Novato, CA, USA) and a microforge (Narishige MF-900, Tokyo, Japan). The outside of the glass capillary tube for the inner fluid was hydrophobically functionalized with OTS to enhance the wettability of the capillary tube with oil phase, facilitating the fabrication of W/O/W double-emulsion drops. The two tapered circular capillary tubes were coaxially aligned and opposed to each other within a square glass capillary (ID of 1.05 mm), which were spaced from each other by 80 μm, as shown the optical microscope image in Figure 2a. All the capillaries and needles used for channels were connected using a transparent epoxy. Such a configuration possessed the hydrodynamic flow-focusing and coflowing functions for one-step fabrication of highly monodisperse W/O/W double-emulsion drops.

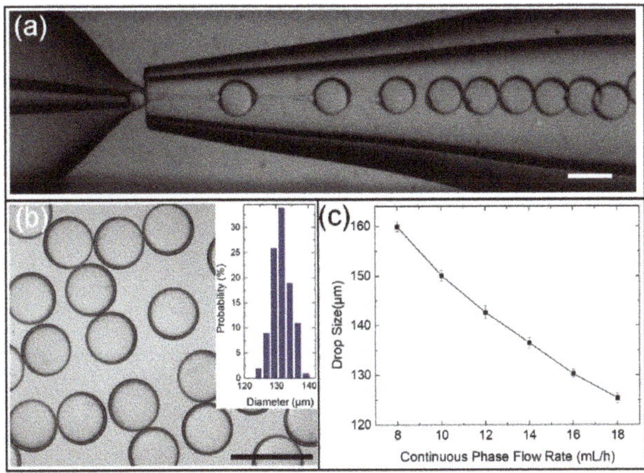

Figure 2. (a) Optical microscope image showing the inner configuration of the classic capillary microfluidic device and the generation of W/O/W double-emulsion drops using this device; (b) optical microscopy image of highly monodisperse W/O/W double-emulsion drops (inset: size distribution of the emulsion drops). (c) Influence of continuous phase flow rate on dimensions of double-emulsion drops, where Q_{outer} is varied from 8 to 18 mL/h; Q_{middle} and Q_{inner} are maintained at 120 and 1000 μL/h. Scale bars are 200 μm.

2.3. Generation of Monodisperse W/O/W Double-Emulsion Drops

To generate W/O/W double-emulsion drops, three liquid phases, W_{outer}, O_{middle} and W_{inner}, were pumped into the microfluidic device by three syringe pumps (Harvard Apparatus, PHD 2000 Series, Holliston, MA, USA). The generation process of the W/O/W double-emulsion drops in the device was monitored by a digital CMOS camera (Prime 95B, Qimaging, Surrey, BC, Canada). By tuning the flow rates of three phases, monodisperse W/O/W double-emulsion drops were formed through one-step emulsification, as shown in Figure 2a, where the flow rates of outer continuous phase (Q_{outer}), middle oil phase (Q_{middle}) and inner aqueous phases (Q_{inner}) are 16,000, 120 and 1000 µL/h, respectively. The double-emulsion drops generated using this classic capillary microfluidic device were highly monodisperse, of which the size distribution can be maintained at 2.35% or less, as shown in Figure 2b and the inset bar graph. The drop diameter (d) was controlled precisely by varying Q_{outer}. For example, as Q_{outer} increased from 8 to 18 mL/h, d decreasedfrom 159.9 to 125.4 µm, as shown in the graph in Fig. 2c, where Q_{middle} and Q_{inner} are maintained at 120 and 1000 µL/h, respectively. The reason can be understood that the increased Q_{outer} induces a higher shear force acting on the emulsion drop, leading to the decrease in d.

In addition, the shell thickness (h) of double-emulsion drops was varied by modulating the flow rates ratios of Q_{middle}/Q_{inner}, as shown in the graph and inset images in Figure 3. The stars represent the experimental results of relative shell thickness $2h/d$, which could be varied from 0.041 up to 0.386 with excellent control and reproducibility. The inset optical microscope images show emulsion drops with varying relative shell thickness: 0.041, 0.128, 0.196, 0.232 and 0.386. Table 1 shows the W/O/W double-emulsion drops with varying relative shell thickness, diameter, shell thickness for the relevant flow ratios of Q_{middle}/Q_{inner}. The line in Figure 3 is plotted to illustrate the relative shell thickness $2h/d$ as a function of Q_{middle}/Q_{inner}, which is rearranged by the mass balance equation in reference [16,45,46]. The experimental results are in good agreement with the line plotted in Figure 3, indicating that the microfluidic emulsion technology is good at controlling the shell thickness of the double-emulsion drops. However, it is difficult to generate monodisperse double-emulsion drops with ultra-thin shells using this classic microfluidic emulsion technology. Based on the reliable generation of monodisperse double-emulsion drops, the osmolarity-controlled swelling method was developed to produce the ultra-thin-shell double-emulsion drops and the resultant microcapsules.

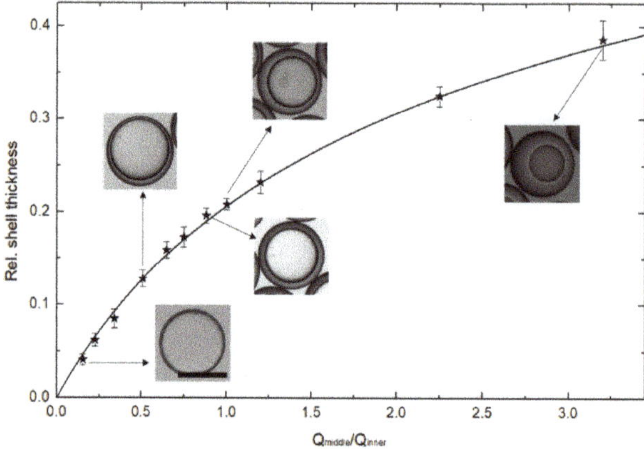

Figure 3. Experimental and theoretic shell thickness as a function of the flow rates ratios of the middle oil and inner aqueous phases. The inset optical microscope images showing emulsion drops with varying relative shell thickness: 0.041, 0.128, 0.196, 0.232 and 0.386. Scale bars are 100 µm.

Table 1. W/O/W double-emulsion drops with varying relative shell thickness, diameter, shell thickness for the relevant flow ratios of Q_{middle}/Q_{inner}.

Q_{middle}/Q_{inner}	Rel. Shell Thickness	D (µm)	h (µm)
0.154	0.041	132.5	2.7
0.511	0.128	139.6	8.9
0.881	0.196	136.6	13.4
1.00	0.208	133.5	13.9
3.20	0.386	132.0	25.5

2.4. Electrotriggered Rupturing of W/O/W Double-Emulsion Drops

In this study, the W/O/W double-emulsion drop needed to be triggered to rupture by applying an alternating current (AC) electric field to calculate the shell thickness. Two acupuncture needles immersed in the suspending medium acted as electrodes with a distance of 800 µm, through which a square-wave AC voltage signal was applied to develop an AC electric field between the electrodes. During this process, the AC signal was generated by a function generator (TGA 12104, TTi, Manchester, UK), and amplified by an amplifier (model 2350, TEGAM, Geneva, OH, USA). Under an appropriate field strength and frequency (30 V, 5 KHz) applied, the double-emulsion drop between the two electrodes ruptured immediately due to Maxwell–Wagner interfacial polarization [47–49].

3. Results and Discussion

3.1. Osmolarity-Controlled Swelling Behavior

The core diameter and shell thickness of the W/O/W double-emulsion drops can be tuned by the osmolarity-controlled swelling behavior. When the salt concentration of outer suspending medium is higher than it in the inner aqueous core, water in the suspending medium has a higher chemical potential than that of the inner core, as shown in the schematic illustration in Figure 4a. As a result, water in the suspending medium will penetrate through the oil shell into the core, leading to the osmolality-controlled swelling behaviors of the inner core [43,50]. Along with the swelling of the inner core, the oil shell becomes sufficiently thinner and thinner, leading to the generation of ultra-thin-shell double-emulsion drops.

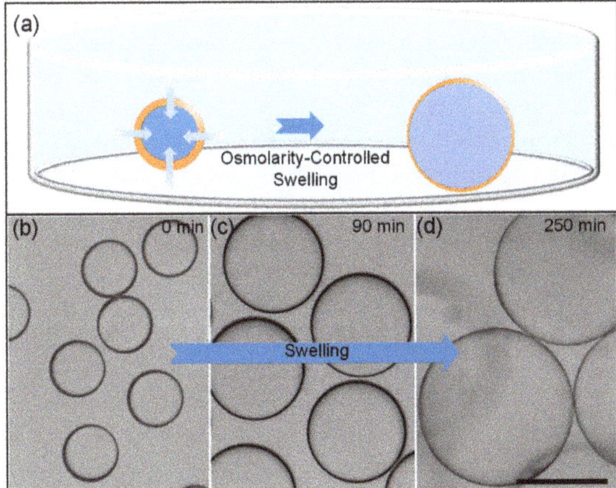

Figure 4. (**a**) Schematic illustration showing the osmolarity-controlled swelling process of the W/O/W double-emulsion drops; (**b–d**) Optical microscope images showing the osmolarity-controlled swelling process of W/O/W double-emulsion. Scale bars are 200 µm.

To study the effect of osmolality-controlled swelling behavior on shell thickness of double-emulsion drops, the swelling phenomena of emulsion drops with varying osmotic pressure between the inner core and outer suspending medium are investigated. Firstly, double-emulsion drops with a given KCl concentration of 0.25 mol/L as the inner aqueous cores are generated in our capillary device, which are collected in 1 wt% PVA solution with KCl concentration of 0.25 mol/L to prevent the spontaneous swelling and coalescence of these drops, as shown in Figure 4b. Here, Q_{outer}, Q_{middle} and Q_{inner} are 16,000, 120 and 1000 μL/h, respectively. The drop diameter is about 129.5 μm. According to the relationship between the relative shell thickness $2h/d$ and Q_{middle}/Q_{inner} shown in Figure 3, we can derive the shell thickness of the double-emulsion drop is 2.4 μm. Then, the KCl concentration of the suspending medium is double diluted by adding 1 wt% PVA solution every 30 min. Here, the dilution of the KCl concentration in the suspending medium induces the varying osmotic pressure between inner cores and the suspending medium, resulting in the osmolality-controlled swelling of the inner core. Meanwhile, the drop size is measured before every dilution process, as shown in the optical images in Figure 4b–d. After five times dilution, finally, the diameter of the double-emulsion drops increases from 129.5 to 360 μm along with the osmolality-controlled swelling process. Knowing the initial diameter, shell thickness and the swelled diameter of the double-emulsion, the shell thickness of the swelled emulsion can be derived geometrically, the value of which is 300 nm. As a result, the shell thickness of the emulsion decreases from 2.4 μm to about 300 nm, indicating that the osmolarity-controlled swelling method enables the mass production of microcapsules with ultra-thin-shell.

3.2. Measuring of the Shell Thickness

To evaluate the shell thickness of the double-emulsion drop after swelling, scanning electron microscope (SEM) imaging is the most direct way to get this value [8,36,40]. However, it needs expensive SEM equipment and multiple operations to take SEM images. In other way, if we know the initial data (diameter and shell thickness before swelling) and the swelled diameter of the double-emulsion drop, the shell thickness of the swelled emulsion drop can be derived geometrically, as demonstrated in Section 3.1. In some cases, however, the initial data of the original capsules are missing or not attainable, leading to the difficulty in calculating shell thickness. Here, we present a simple and useful strategy to quickly derive the shell thickness by measuring the volume of oil shell phase after the emulsion drop rupturing. As a comparison, the original emulsion drops are generated using the same flow rates with Section 3.1, the diameter and shell thickness of which are 129.5 and 2.4 μm, respectively. Time-lapse microscopy images in Figure 5 show the rupture process. Methylene blue is dissolved into the inner core as a dye for better visualization. When the emulsion drop is swelling to about 360 μm in diameter, the drop is triggered to rupture by applying an alternating electric field as described in Section 2.4. From 0.5 s, the aqueous core of the emulsion drop sprays out because of the interfacial tension, and the oil shell shrinks into a wrinkled membrane. At about 35 s, most of the inner phase is discharged from the core. The empty oil shell assembles into a spherical drop under the action of surface tension. After 75 s, the oil shell transformed into a tiny oil drop with a diameter of 64 μm. By comparing the diameters of the oil drop after rupturing and the double-emulsion drop before rupturing, we can readily figure out the shell thickness of the swelled double-emulsion drop, which is about 338 nm. This indicates that the shell thickness evaluated using the drop rupturing method (338 nm) is in good agreement with that derived geometrically (300 nm).

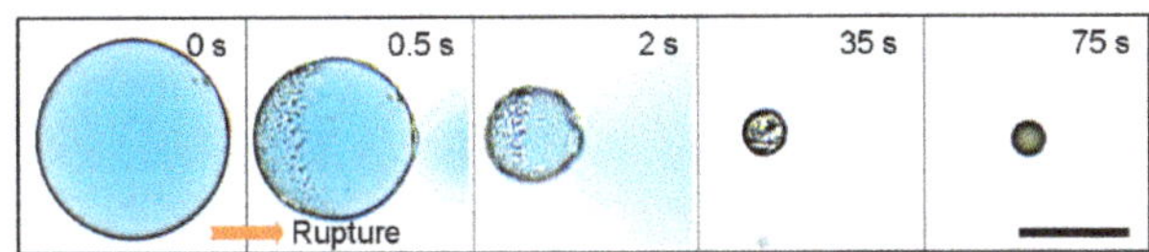

Figure 5. Time-lapse microscopy images showing the rupture process of the W/O/W double-emulsion drop. A digital charge coupled device (CCD) camera (Retiga 2000R, Qimaging) is used to take these images. Scale bar is 200 μm.

3.3. Curing Process – Transform the Double-Emulsion Drops to Solid Core-Shell Microcapsules

For some situations, the emulsion drops need to be transferred to microcapsules for long-term observation and thin shell buckling investigation [35]. Herein, we cure the PDMS shell at 37 °C for 24 h to obtain the microcapsules with highly robust ultra-thin shells, as the schematic illustration shows in Figure 6a. An number of solid microcapsules with an ultra-thin shell after thermal curing is shown in Figure 6b. As can be seen in Figures 2b and 6b, the size distribution of the double-emulsion drops (polydispersity of 2.35%) and swelling produced ultra-thin-shell microcapsules (polydispersity of 2.56%) shows good uniformity, clearly indicating that the osmolality-controlled swelling method for mass production of ultra-thin-shell microcapsules is feasible and efficient. Furthermore, the solid ultra-thin-shell microcapsules can maintain their intact spherical shape for up to 1 year with no rupture in our long-term observation, indicating that the ultra-thin-shell PDMS microcapsules are robust as vesicles for long-term encapsulation and observation.

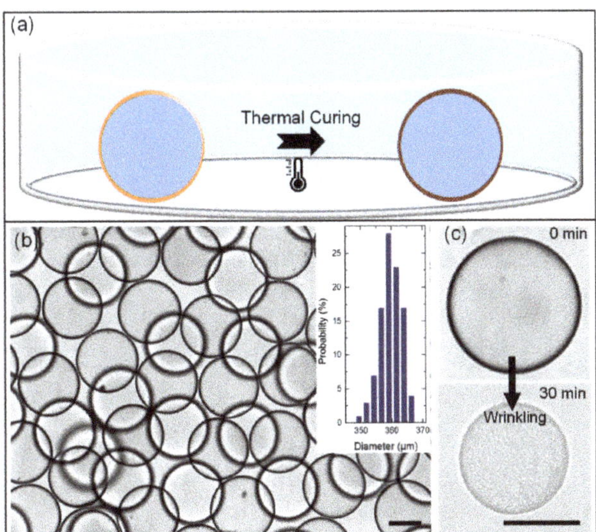

Figure 6. (a) Schematic illustration showing the liquid emulsion drop transformed to solid microcapsule via the thermal curing process; (b) solid microcapsules with ultra-thin shell after the thermal curing process (inset: size distribution of the microcapsules); (c) wrinkling of cured microcapsules by imposing a strong osmotic pressure. Scale bars are 200 μm.

To demonstrate the uniformity of the shell thickness, we compress the ultra-thin-shell microcapsules by imposing a strong osmotic pressure. When these ultra-thin-shell microcapsules are placed in a suspending medium with KCl concentration of 0.5 mol/L, the inner aqueous phase diffuses out from the PDMS microcapsules, resulting in a volume shrinkage of the microcapsules. As the shrink process goes on, wrinkles develop on the ultra-thin-shell microcapsules. After 30 min, wrinkles with remarkable uniformity grow up on the shell of microcapsule, as shown in Figure 6c. Direct observation of uniform wrinkling on the surface of the ultra-thin-shell microcapsules after osmolality-controlled shrink confirms that the shell thickness of the microcapsules is ultra-thin and relatively uniform. We believe this kind of ultra-thin-shell PDMS microcapsule is an attractive template for studies of folding, buckling and rupturing behaviors of spherical thin films.

4. Conclusions

In this study, an osmolarity-controlled swelling method is presented to generate highly monodisperse microcapsules with an ultra-thin shell via W/O/W double-emulsion drops templating. The shell thickness of the emulsion drops can be tuned precisely off-chip through osmolarity-induced swelling of the emulsion drops. Here, this swelling process is simply controlled by the osmotic pressure between the inner aqueous and the suspending medium. Using this method, a shell thickness of hundreds of nanometers can be developed. Furthermore, after a thermal-curing process, the emulsion drops can be transformed to ultra-thin-shell microcapsules, which can maintain their intact spherical shape for up to 1 year with no rupture in our long-term observation. This method enables the mass production of microcapsules with ultra-thin shells, which are robust as vesicles for long-term encapsulation and observation, and attractive templates for studies of folding, buckling and rupturing behaviors of spherical thin films.

Author Contributions: J.G. designed the experiment and wrote the original manuscript. L.H. processed the experimental data, J.Y. and J.H. conducted the experiment; Q.H. provided the idea and revised the manuscript. All authors have read and agreed to the published version of the manuscript.

Funding: This work is supported by the University Nursing Program for Young Scholars with Creative Talents in Heilongjiang Province (Grant No. UNPYSCT-2018104), Basic scientific research operating expenses project of Heilongjiang Province (Grant No. 135409604 and 135409102), the Industry-Academic Cooperative Education Project of Ministry of Education (Grant No. 201901145012), and Research Project on Higher Education Reform in Heilongjiang Province (Grant No. SJGY20190715).

Conflicts of Interest: The authors declare no conflict of interest.

References

1. Datta, S.S.; Abbaspourrad, A.; Amstad, E.; Fan, J.; Kim, S.H.; Romanowsky, M.; Shum, H.C.; Sun, B.; Utada, A.S.; Windbergs, M.; et al. 25th anniversary article: Double emulsion templated solid microcapsules: Mechanics and controlled release. *Adv. Mater.* **2014**, *26*, 2205–2218. [CrossRef] [PubMed]

2. Sanchez Barea, J.; Lee, J.; Kang, D.K. Recent Advances in Droplet-based Microfluidic Technologies for Biochemistry and Molecular Biology. *Micromachines (Basel)* **2019**, *10*, 412. [CrossRef] [PubMed]

3. Mai, Z.; Chen, J.; He, T.; Hu, Y.; Dong, X.; Zhang, H.; Huang, W.; Ko, F.; Zhou, W. Electrospray biodegradable microcapsules loaded with curcumin for drug delivery systems with high bioactivity. *RSC Adv.* **2017**, *7*, 1724–1734. [CrossRef]

4. Lee, W.L.; Tan, J.W.; Tan, C.N.; Loo, S.C. Modulating drug release from gastric-floating microcapsules through spray-coating layers. *PLoS ONE* **2014**, *9*, e114284. [CrossRef] [PubMed]

5. You, X.R.; Ju, X.J.; He, F.; Wang, Y.; Liu, Z.; Wang, W.; Xie, R.; Chu, L.Y. Polymersomes with Rapid K(+)-Triggered Drug-Release Behaviors. *ACS Appl. Mater. Interfaces* **2017**, *9*, 19258–19268. [CrossRef] [PubMed]

6. Li, W.; Dong, H.; Tang, G.; Ma, T.; Cao, X. Controllable microfluidic fabrication of Janus and microcapsule particles for drug delivery applications. *RSC Adv.* **2015**, *5*, 23181–23188. [CrossRef]

7. Sun, S.; Liang, N.; Gong, X.; An, W.; Kawashima, Y.; Cui, F.; Yan, P. Multifunctional Composite Microcapsules for Oral Delivery of Insulin. *Int. J. Mol. Sci.* **2016**, *18*, 54. [CrossRef]

8. Kim, S.H.; Park, J.G.; Choi, T.M.; Manoharan, V.N.; Weitz, D.A. Osmotic-pressure-controlled concentration of colloidal particles in thin-shelled capsules. *Nat. Commun.* **2014**, *5*, 3068. [CrossRef]

9. Choi, T.M.; Je, K.; Park, J.G.; Lee, G.H.; Kim, S.H. Photonic Capsule Sensors with Built-In Colloidal Crystallites. *Adv. Mater.* **2018**, *30*, e1803387. [CrossRef]

10. Zhao, Y.; Xie, Z.; Gu, H.; Jin, L.; Zhao, X.; Wang, B.; Gu, Z. Multifunctional photonic crystal barcodes from microfluidics. *NPG Asia Mater.* **2012**, *4*, e25. [CrossRef]

11. Niepa, T.H.; Hou, L.; Jiang, H.; Goulian, M.; Koo, H.; Stebe, K.J.; Lee, D. Microbial Nanoculture as an Artificial Microniche. *Sci. Rep.* **2016**, *6*, 30578. [CrossRef]

12. Chang, C.B.; Wilking, J.N.; Kim, S.H.; Shum, H.C.; Weitz, D.A. Monodisperse Emulsion Drop Microenvironments for Bacterial Biofilm Growth. *Small* **2015**, *11*, 3954–3961. [CrossRef] [PubMed]

13. Lu, W.; Kelly, A.L.; Miao, S. Emulsion-based encapsulation and delivery systems for polyphenols. *Trends Food Sci. Technol.* **2016**, *47*, 1–9. [CrossRef]

14. Madene, A.; Jacquot, M.; Scher, J.; Desobry, S. Flavour encapsulation and controlled release—A review. *Int. J. Food Sci. Technol.* **2006**, *41*, 1–21. [CrossRef]

15. Xie, X.; Zhang, W.; Abbaspourrad, A.; Ahn, J.; Bader, A.; Bose, S.; Vegas, A.; Lin, J.; Tao, J.; Hang, T.; et al. Microfluidic Fabrication of Colloidal Nanomaterials-Encapsulated Microcapsules for Biomolecular Sensing. *Nano Lett.* **2017**, *17*, 2015–2020. [CrossRef]

16. Chen, P.W.; Erb, R.M.; Studart, A.R. Designer polymer-based microcapsules made using microfluidics. *Langmuir* **2012**, *28*, 144–152. [CrossRef] [PubMed]

17. Utada, A.S.; Lorenceau, E.; Link, D.R.; Kaplan, P.D.; Stone, H.A.; Weitz, D.A. Monodisperse double emulsions generated from a microcapillary device. *Science* **2005**, *308*, 537–541. [CrossRef]

18. Goertz, J.P.; DeMella, K.C.; Thompson, B.R.; White, I.M.; Raghavan, S.R. Responsive capsules that enable hermetic encapsulation of contents and their thermally triggered burst-release. *Mater. Horiz.* **2019**, *6*, 1238–1243. [CrossRef]

19. Grolman, J.M.; Inci, B.; Moore, J.S. pH-Dependent Switchable Permeability from Core–Shell Microcapsules. *ACS Macro Lett.* **2015**, *4*, 441–445. [CrossRef]

20. Zhou, S.; Fan, J.; Datta, S.S.; Guo, M.; Guo, X.; Weitz, D.A. Thermally Switched Release from Nanoparticle Colloidosomes. *Adv. Funct. Mater.* **2013**, *23*, 5925–5929. [CrossRef]

21. Windbergs, M.; Zhao, Y.; Heyman, J.; Weitz, D.A. Biodegradable core-shell carriers for simultaneous encapsulation of synergistic actives. *J. Am. Chem. Soc.* **2013**, *135*, 7933–7937. [CrossRef] [PubMed]

22. DiLauro, A.M.; Abbaspourrad, A.; Weitz, D.A.; Phillips, S.T. Stimuli-Responsive Core–Shell Microcapsules with Tunable Rates of Release by Using a Depolymerizable Poly(phthalaldehyde) Membrane. *Macromolecules* **2013**, *46*, 3309–3313. [CrossRef]

23. Abbaspourrad, A.; Datta, S.S.; Weitz, D.A. Controlling release from pH-responsive microcapsules. *Langmuir* **2013**, *29*, 12697–12702. [CrossRef]

24. Abbaspourrad, A.; Carroll, N.J.; Kim, S.H.; Weitz, D.A. Polymer microcapsules with programmable active release. *J. Am. Chem. Soc.* **2013**, *135*, 7744–7750. [CrossRef]

25. Amstad, E.; Kim, S.H.; Weitz, D.A. Photo- and thermoresponsive polymersomes for triggered release. *Angew. Chem. Int. Ed. Engl.* **2012**, *51*, 12499–12503. [CrossRef] [PubMed]

26. Zhang, X.P.; Luo, J.; Zhang, D.X.; Jing, T.F.; Li, B.X.; Liu, F. Porous microcapsules with tunable pore sizes provide easily controllable release and bioactivity. *J. Colloid Interface Sci.* **2018**, *517*, 86–92. [CrossRef] [PubMed]

27. Mou, C.-L.; Wang, W.; Li, Z.-L.; Ju, X.-J.; Xie, R.; Deng, N.-N.; Wei, J.; Liu, Z.; Chu, L.-Y. Trojan-Horse-Like Stimuli-Responsive Microcapsules. *Adv. Sci.* **2018**, *5*, 1700960. [CrossRef] [PubMed]

28. Lee, S.; Lee, T.Y.; Amstad, E.; Kim, S.H. Microfluidic Production of Capsules-in-Capsules for Programed Release of Multiple Ingredients. *Adv. Mater. Technol.-US* **2018**, *3*, 1800006. [CrossRef]

29. Sun, B.J.; Shum, H.C.; Holtze, C.; Weitz, D.A. Microfluidic melt emulsification for encapsulation and release of actives. *ACS Appl. Mater. Interfaces* **2010**, *2*, 3411–3416. [CrossRef]

30. Wang, W.; Luo, T.; Ju, X.J.; Xie, R.; Liu, L.; Chu, L.Y. Microfluidic Preparation of Multicompartment Microcapsules for Isolated Co-encapsulation and Controlled Release of Diverse Components. *Int. J. Nonlinear Sci. Numer. Simul.* **2012**, *13*, 325–332. [CrossRef]

31. Kim, S.H.; Kim, J.W.; Kim, D.H.; Han, S.H.; Weitz, D.A. Polymersomes containing a hydrogel network for high stability and controlled release. *Small* **2013**, *9*, 124–131. [CrossRef] [PubMed]

32. Liu, Y.-M.; Wu, W.; Ju, X.-J.; Wang, W.; Xie, R.; Mou, C.-L.; Zheng, W.-C.; Liu, Z.; Chu, L.-Y. Smart microcapsules for direction-specific burst release of hydrophobic drugs. *RSC Adv.* **2014**, *4*, 46568–46575. [CrossRef]

33. Huang, W.; Chen, Y.; Chen, L.; Zhong, J.; Johri, A.M.; Zhou, J. Multimodality imaging-guided local injection of eccentric magnetic microcapsules with electromagnetically controlled drug release. *Cancer Rep.* **2018**, *2*, e1154. [CrossRef]

34. Huang, J.; Luo, C.; Li, W.; Li, Y.; Zhang, Y.S.; Zhou, J.; Jiang, Q. Eccentric magnetic microcapsules for orientation-specific and dual stimuli-responsive drug release. *J. Mater. Chem. B* **2015**, *3*, 4530–4538. [CrossRef]

35. Katifori, E.; Alben, S.; Cerda, E.; Nelson, D.R.; Dumais, J. Foldable structures and the natural design of pollen grains. *Proc. Natl. Acad. Sci. USA* **2010**, *107*, 7635–7639. [CrossRef]

36. Kim, S.H.; Kim, J.W.; Cho, J.C.; Weitz, D.A. Double-emulsion drops with ultra-thin shells for capsule templates. *Lab Chip* **2011**, *11*, 3162–3166. [CrossRef]

37. Arriaga, L.R.; Datta, S.S.; Kim, S.H.; Amstad, E.; Kodger, T.E.; Monroy, F.; Weitz, D.A. Ultrathin shell double emulsion templated giant unilamellar lipid vesicles with controlled microdomain formation. *Small* **2014**, *10*, 950–956. [CrossRef]

38. Zhao, C.X.; Chen, D.; Hui, Y.; Weitz, D.A.; Middelberg, A.P.J. Controlled Generation of Ultrathin-Shell Double Emulsions and Studies on Their Stability. *Chemphyschem* **2017**, *18*, 1393–1399. [CrossRef]

39. Zhang, W.X.; Abbaspourrad, A.; Chen, D.; Campbell, E.; Zhao, H.; Li, Y.W.; Li, Q.N.; Weitz, D.A. Osmotic Pressure Triggered Rapid Release of Encapsulated Enzymes with Enhanced Activity. *Adv. Funct. Mater.* **2017**, *27*, 1700975. [CrossRef]

40. Lee, T.Y.; Ku, M.; Kim, B.; Lee, S.; Yang, J.; Kim, S.H. Microfluidic Production of Biodegradable Microcapsules for Sustained Release of Hydrophilic Actives. *Small* **2017**, *13*, 1700646. [CrossRef]

41. Kim, B.; Jeon, T.Y.; Oh, Y.K.; Kim, S.H. Microfluidic Production of Semipermeable Microcapsules by Polymerization-Induced Phase Separation. *Langmuir* **2015**, *31*, 6027–6034. [CrossRef] [PubMed]

42. Chaurasia, A.S.; Josephides, D.N.; Sajjadi, S. Large ultrathin shelled drops produced via non-confined microfluidics. *Chemphyschem* **2015**, *16*, 403–411. [CrossRef] [PubMed]

43. Guan, X.; Hou, L.; Ren, Y.; Deng, X.; Lang, Q.; Jia, Y.; Hu, Q.; Tao, Y.; Liu, J.; Jiang, H. A dual-core double emulsion platform for osmolarity-controlled microreactor triggered by coalescence of encapsulated droplets. *Biomicrofluidics* **2016**, *10*, 034111. [CrossRef] [PubMed]

44. Hou, L.; Ren, Y.; Jia, Y.; Chen, X.; Deng, X.; Tang, Z.; Hu, Q.; Tao, Y.; Jiang, H. Osmolarity-controlled swelling behaviors of dual-cored double-emulsion drops. *Microfluid. Nanofluid.* **2017**, *21*, 60. [CrossRef]

45. Lee, D.; Weitz, D.A. Double Emulsion-Templated Nanoparticle Colloidosomes with Selective Permeability. *Adv. Mater.* **2008**, *20*, 3498–3503. [CrossRef]

46. Hennequin, Y.; Pannacci, N.; de Torres, C.P.; Tetradis-Meris, G.; Chapuliot, S.; Bouchaud, E.; Tabeling, P. Synthesizing microcapsules with controlled geometrical and mechanical properties with microfluidic double emulsion technology. *Langmuir* **2009**, *25*, 7857–7861. [CrossRef] [PubMed]

47. Jia, Y.; Ren, Y.; Hou, L.; Liu, W.; Jiang, T.; Deng, X.; Tao, Y.; Jiang, H. Electrically controlled rapid release of actives encapsulated in double-emulsion droplets. *Lab Chip* **2018**, *18*, 1121–1129. [CrossRef] [PubMed]

48. Deng, X.; Ren, Y.; Hou, L.; Liu, W.; Jia, Y.; Jiang, H. Electric Field-Induced Cutting of Hydrogel Microfibers with Precise Length Control for Micromotors and Building Blocks. *ACS Appl. Mater. Interfaces* **2018**, *10*, 40228–40237. [CrossRef] [PubMed]

49. Deng, X.; Ren, Y.; Hou, L.; Liu, W.; Jiang, T.; Jiang, H. Compound-Droplet-Pairs-Filled Hydrogel Microfiber for Electric-Field-Induced Selective Release. *Small* **2019**, *15*, e1903098. [CrossRef]

50. Tu, F.; Lee, D. Controlling the stability and size of double-emulsion-templated poly(lactic-co-glycolic) acid microcapsules. *Langmuir* **2012**, *28*, 9944–9952. [CrossRef]

Communication

Air-Bubble Induced Mixing: A Fluidic Mixer Chip

Xiaoyu Jia [1,†], Bingchen Che [2,†], Guangyin Jing [1,2,*] and Ce Zhang [2,*]

1   School of Physics, Northwest University, Xi'an 710069, China; 201720556@stumail.nwu.edu.cn
2   State Key Laboratory of Cultivation Base for Photoelectric Technology and Functional Materials, Institute of Photonics and Photon-Technology, Northwest University, Xi'an 710069, China; bingchenc03@gmail.com
*   Correspondence: jing@nwu.edu.cn (G.J.); zhangce@nwu.edu.cn (C.Z.);
    Tel.: +86-29-88303281 (G.J.); +86-29-88303281 (C.Z.)
†   These authors contribute equally to this work.

Received: 3 January 2020; Accepted: 11 February 2020; Published: 14 February 2020

Abstract: In this study, we report the design and fabrication of a novel fluidic mixer. As proof-of-concept, the laminar flow in the main channel is firstly filled with small air-bubbles, which act as active stirrers inducing chaotic convective turbulent flow, and thus enhance the solutes mixing even at a low input flow rate. To further increase mixing efficiency, a design of neck constriction is included, which changes the relative positions of the inclusion bubbles significantly. The redistribution of liquid volume among bubbles then causes complex flow profile, which further enhances mixing. This work demonstrates a unique approach of utilizing air bubbles to facilitate mixing in bulk solution, which can find the potential applications in microfluidics, fast medical analysis, and biochemical synthesis.

Keywords: fluidic mixer; air bubble; 3D printing

1. Introduction

Mixing of the solutes in fluid with an efficient way is essential in chemical, biological, medical, and material industries [1–6]. A fluidic mixer is not only a crucial component for lab-on-a-chip studies, but also has great practical applications [7,8]. The uniform laminar flow is the characteristic feature of the fluidic channels, which use basic diffusion process plus the sharp wedge for the passive slow mixing [9]. For practical use, it is highly desirable to mix the solute across the channel section before the solute is transported forward. To accomplish fully chaotic mixing, there emerged many fluidic mixers over the years, which can be cataloged into two types: the passive and the active mixer. The passive mixer utilizes geometric obstacles to create the flow field perturbation [10–14], such as intersection channels [15,16], convergent–divergent channels [17], three-dimensional architectures by multiple fabrications of the channels [18,19], and embedded, barrier-based obstacles [20]. Curved wall of the channel is better to disturb the flow than the flat straight one. Periodic wavy poly-dimethylsiloxane (PDMS) channels was designed to enhance the mixing process due to the vortex generation located at the higher curved part [21]. The active fluidic mixer induces external energy sources, which can be mechanical pulsation or electrokinetic forces [22–26]. Even though active fluidic mixer is favorable due to many adjustable parameters (e.g., perturbation frequency, phase, and amplitude), only a few studies have been reported due to the complexity of control schemes and difficulties in fabrications. The droplet mixer generates chaotic advection by pressure-driving a flow cavity (droplet) through fluidic channels of various geometries [27,28]. Relative motion of the surrounding walls induces convection flows that mix solutions. As a passive mixer, the mixing mostly depends on the geometry of fluidic channels, which determines whether the solute can pass across the streamlines. Therefore, there may still be islands within the flow cavity, where the solute is not touchable. A combination of channel geometry and the inclusion of a slip boundary into the flow, i.e., emulsions or bubbly flow, is believed to be demanding for mixing in microfluidic channels. We herein present a simple design, using air-bubbles as the stirrer. We propose that the presence of bubbles within the otherwise laminar flow can stimulate a fascinating variety of motion patterns on one hand, and the instabilities to transfer fluid momentum with the uniformization process on the other hand.

2. Fabrication and Experiments

2.1. Chip Fabrication

Chip Fabrication was performed using the standard soft lithography approach in combination with 3D printing (Figure S1) [29]. The poly-dimethylsiloxane (PDMS) device can consist of one or multiple layers, depending on the chip functionality. The 3D-printer (CR-3040, CREALITY, Shenzhen, China), which can produce structures with a minimum width of 0.4 mm and a minimum height of 0.1 mm, was used to create negative templates for each layer. The 3D-printed template surface was smoothed with alcohol before PDMS casting. To fabricate fluidic chips with only one layer, the 3D-printed mold was cast with a PDMS mixture with a curing agent at 5:1, and cured by incubating at 45 °C for 1 h to 2 h depending on the thickness. Subsequently, the resulting PDMS duplicated get holes punched where the gas and solution inputs were connected (see Figure 1). The fluidic device was sealed using glass slides, which were spin-coated with a 20:1 PDMS mixer at 1200 rpm and cured at 80 °C for 1 h. Following standard PDMS-PDMS bonding protocol (i.e., off-ratio bonding), the PDMS devices (5:1) and PDMS coated glass slides (20:1) were brought in contact and incubated at 80 °C for 24 h before use. For multilayer fluidic devices (e.g., the one shown in Figure S2), the same protocol applied with alternative 20:1 and 5:1 mixing ratios starting from the top flow layer.

2.2. Chip Control and Operation

To set up a chip, we connected the PDMS chips to tubes filled with DI-water via TYGON tubing (Milan, Satigny, Switzerland). In order to degas the chip, the outlet was firstly sealed and air was fully pressed out through the porous PDMS material, while fluid remained in the chip. The process continued till the flow layer was fully filled by DI water. During experiments, 4 inlets were connected to food dyes of different colors, and one inlet was connected to a pressurized gas pipeline.

2.3. Data Analysis

To quantify the color transition during mixing, we split the optical images into red, green and blue mono-color channels. We picked the red channel, in which the intensity difference reflects the color transitions during mixing. Intensity variances at different positions along the fluidic channel were collected, and plotted against the distance away from the inlet, as illustrated in Figure S3.

3. Results and Discussion

3.1. The Theory

The key feature of the proposed fluidic mixer is to generate air-bubbles within bulk solutions in a controllable manner. The size and number of the bubbles in the continuous flow are set by the relative flow rate between liquid phase and gas phase. Basically, the movement of bubbles in a straight PDMS-based fluidic channel depends on the number density of the bubbles, their sizes, their shapes, and their positions in the channel, which vary greatly among neighboring bubbles. The constant movements of bubbles with respect to one another create ever-changing liquid volume redistribution, and thus cause turbulent flow, favoring fast mixing. Suppose the simplest case is to solve the Navier-Stockes equation for lateral velocity v_y in the flow (perpendicular to the flow direction) with the boundary Ω between the continuous flow phase and one individual bubble.

$$\frac{\partial v_y}{\partial t} + v_x \frac{\partial v_y}{\partial x} + v_y \frac{\partial v_y}{\partial y} = -\frac{1}{\rho}\frac{\partial p}{\partial y} + v\left(\frac{\partial^2 v_y}{\partial x^2} + \frac{\partial^2 v_y}{\partial y^2}\right) \tag{1}$$

With the coupling of the boundary constriction in the normal and tangential components $\vec{v}\,|_\Omega = 0$, $\vec{n}\cdot\overline{\overline{\tau}}\,|_\Omega = \kappa\gamma\vec{n}$, plus the incompressible assumption $\frac{\partial v_x}{x} + \frac{\partial v_y}{y} = 0$, it is possible to solve these set of equations numerically based on computational algorithm for the similar situation of bubble rising in stationary liquid [30]. From the experimental point of view, obviously, the lateral convective velocity is enhanced by the introducing of bubbles, compared with that only by the thermal diffusion. The presence of groups of bubbles in the fluid allows for chaotic flow with the benefit for mixing. The bubbles change position, shape, and size, and therefore cause bubble swarming. To further enhance mixing, here we experimentally include the constrictions of various sizes into the design of the channel geometry, which makes it difficult to solve the analytics, even numerically, for the velocity field.

For instance, the interaction between bubbles and walls is tuned by the complex friction, and coupled deformation among bubbles by convergent-divergent constriction. Particularly, the deformation controlled by the competition of surface tension and velocity gradient makes it complex to find the detailed boundary, which involves solving the governing equation. However, from the competition among the viscous forces, inertia forces, and capillary forces in the bubbly flow, we can demonstrate how they determine the mixing in the channels.

3.2. The Fluidic Mixer

Based on above-mentioned conception, we fabricate a fluidic chip using 3D printing and the soft lithography techniques, which consists of only one poly-dimethylsioxane (PDMS) layer. The key elements of the fluidic mixer are the varying constriction dimension and a Y-shaped conjunction, where air bubbles are frequently generated (see Figure 1). The Y-channels are 0.5 mm deep and 1 mm wide with two branches connected to the solutes and air, respectively. With a constant input flow rate using a syringe pump, the air pressure determines the size of air-bubbles (Supplementary Section S1). As the generation of droplets can be disturbed by numerous factors (e.g., tube misplacement and deformation of PDMS channels when being pressurized), the bubble size varies randomly even at constant air-pressure. To create physical obstacles, the neck constriction is made 2 mm (wide) by 1 mm (deep) for the wide channels, and 0.5 mm by 1 mm for narrow parts.

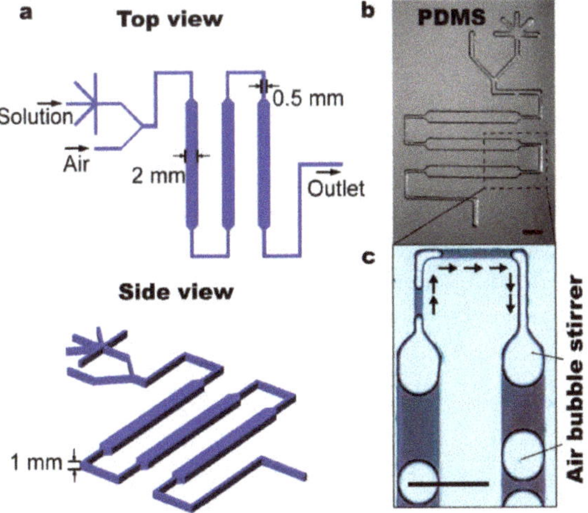

Figure 1. Schematic drawing (**a**) and optical image (**b**) of the fluidic mixer. (**c**) Air bubbles generated on-chip and transport within fluidic channels. Scale bars denote 4 mm in all figures.

3.3. Recirculating Flow in the Moving Droplet

It is demonstrated that with no air input, solutes of the five inputs remain separated near the end of the fluidic chip (Figure 2a). The gradual blurry in the boundary from positions (1) to (3) can only be attributed to thermal dynamic diffusion and minor turbulent flow at the corner section. At high input air pressure, laminar streams of five solutions were frequently separated by air input and form individual droplets (Figure 2b). The mixing efficiency is satisfactory. Different coloring at the top and bottom parts of the droplet is still visible at position (1), which is ≈1 cm away from the Y-junction and position (2) (≈3 cm to the Y-junction). Apparently, it requires a certain distance for the droplet to move, which ensures thorough mixing. The flow profile is traced using fluorescent particles of ≈1 µm in diameter. We observe that the droplets moving in the straight parts of the fluidic channels generate a steady recirculating flow (Figure 3a–c) [31]. Even though the velocity perpendicular to fluidic channel can be as high as two times the driving flow rate, the direction distribution is symmetric, suggesting circulating flow within only half of the droplet (Figure 3d and Video S1). The flow rate along the channel direction is mostly slower than the driving flow rate. As reported before, the 3D flow allows mixing within half of the droplet, and thus the mass transfer in the transversal direction still relies on molecular diffusion and takes time [27,31]. The changed dimension of constrictions along the fluidic channel induces chaotic advection in individual droplets by dynamically changing the shape of the droplet and introducing asymmetry into the recirculating flow system [27]. As the convection flow rate is determined by droplet moving speed, the mixing efficiency depends strongly on the driving flow rate and channel geometry.

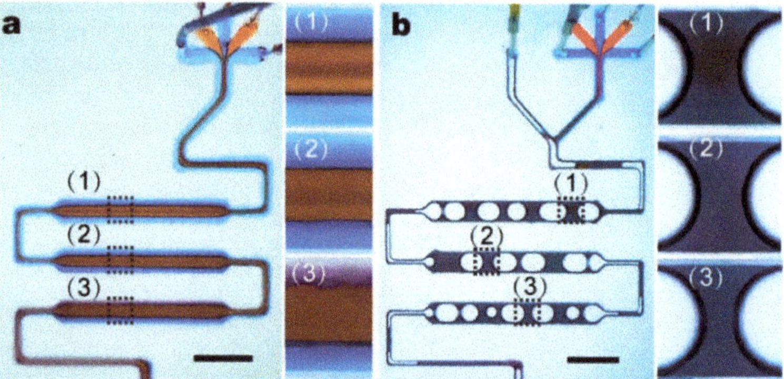

Figure 2. Comparison of mixing efficiency using food dyes between fluidic mixer with (**a**) laminar flow and (**b**) air-bubbles as stirrer. It is demonstrated that with air bubbles, food dyes of different colors can be quickly mixed. Scale bars denote 5 mm in all figures.

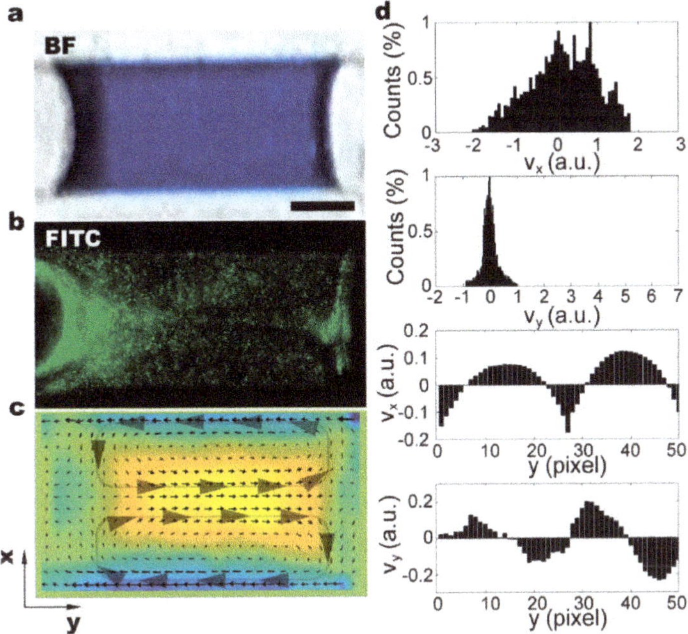

Figure 3. (**a–c**) 3D flow generated within the droplet in straight channels of the chip. It is demonstrated that circulating flow occurs within half of the droplet, and the mass transfer in the transversal direction still relies on molecular diffusion. (**d**) Distribution of flow rate within the moving droplet. The flow velocity is normalized to the driving flow rate to illustrate the mixing efficiency. We observe that flow distribution is symmetrical within a droplet, showing opposite flow profiles. The scale bar denotes 200 μm.

3.4. Small Air-Bubbles as Stirrers

Small bubbles with diameters ranging from 0.1 to 0.3 mm are generated in the fluidic channels at low air pressure (0.05 to 0.1 psi) (Figure 4). For each experiment, the input air-pressure is maintained at a constant value. As mentioned before, the variances in droplet size can be caused by minor disturbances from the environment. As the height of the free droplet is considerably higher than the fluidic channel (i.e., 1 mm), the droplets show an elliptical shape when being confined in the channel, resulting in large contact surface between bubbles and the channel walls. Like stranding pillars, these small bubbles firstly create physical obstacles for the laminar flow (Figure 4a), directing stream lines in the cross-sectional direction through constrictions (Figure 4b). The perturbed laminar flow thus enhances mixing capacities by generating stronger shear and more frequent chaotic advection (Figure 4c).

Notably, these small bubbles move with the bulk fluid (Video S2). When passing through constrictions, strong shear generated at the sides further facilitates mixing, preventing the fluidic channels from being fully blocked. Additionally, bubbles regain their round-shaped conformation when being "pushed" out (Figure 4e). The sudden changes in bubble shape and confinement space create chaotic advection, and facilitate mixing.

When groups of bubbles transporting in fluidic channels, there are three forces at play: viscous forces, inertia forces, and capillary forces. The ratio of any two gives three dimensionless numbers: Reynolds number R_e, comparing inertia to viscos forces; capillary number C_a, comparing viscous to capillary forces; and Weber number W_e, comparing inertia to capillary forces. The detailed expressions are shown below,

$$Re = \rho U L / \mu \tag{2}$$

$$Ca = \mu U / \gamma \tag{3}$$

$$We = \rho U^2 L / \gamma \tag{4}$$

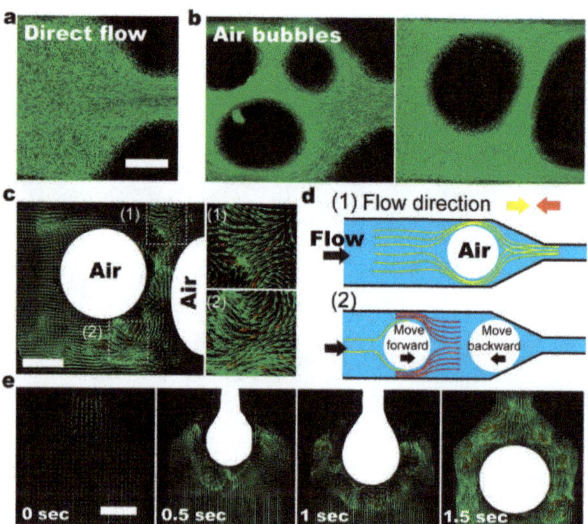

Figure 4. (**a,b**) Flow profile within the fluidic channels using air-bubbles (dark-circles) as the stirrer. Fluorescent particles of ≈1 mm in diameter are used as a flow tracer. (**c**) The flow profile around two air bubbles suggests turbulent flow and mass fluid transportation. (**d**) Schematic shows the mechanism of air bubbles working as the stirrer. (1) In the left panel, the existence of air bubble changes the channel cross-sectional area, and thus induces strong shear around the bubble. (2) The movement of air bubbles can be stopped by changed constriction geometry along the way. Meanwhile, the motion of the following bubbles remains, which changes the relative positions between neighboring bubbles. Consequently, liquid volume redistribution causes turbulent flow, and thus, fast mixing. (**e**) Time series showing the turbulent flow generated when an air bubble moves out from a narrow channel. Scale bars denote 500 µm in all figures.

The typical bubble size is around one millimeter in diameter, which gives the characteristic length $L \approx 1$ mm. The typical velocity of bubbles is around millimeter per second, which gives the characteristic velocity $U \approx 1$ mm/s. The values of density, viscosity, and surface tension are taken from an aqueous solution, leading to $R_e \approx 1$, $C_a \approx 10^{-4}$–10^{-5}, and $W_e \approx 10^{-4}$–10^{-5}.

The bubbles are deformed by the competition between viscous driving forces enacted by surrounding fluids and resistant surface tension. The transverse velocities are caused by the inclusions, deformations, and slip boundaries of bubbles. A group of bubbles in the fluid lead to chaotic flow, benefiting mixing. The varying of bubble positions, shapes, and sizes can therefore cause complex bubble swarming. Note that the interaction between bubbles and channel walls in direct contact, due to the friction, slows down or even trap bubbles, leading to the deformation of streamlines. Besides, the diverging part of the channel creating a compressive flow results in an instability of particles' positions downstream.

Consequently, the transportation speeds of air-bubbles in a fluidic channel are different even when they are similar in size. The altered relative positions of air-bubbles cause liquid volume redistribution, and thus, chaotic advection (Figure 5a and Video S2). It is demonstrated that within a selected area, the velocity perpendicular to fluidic channel can be as large as two times the driving flow rate, which is on par with the recirculating flow in the moving droplet (Figure 5b). The flow distribution, however, is highly asymmetrical, suggesting directional flow across the whole channel. Intriguingly, the velocity along the channels can be four times larger than the driving flow rate, which is considerably higher than both the laminar flow and droplet mixing. Therefore, we conclude that the flow distribution with small bubbles as stirrers is ever-changing and mostly random. Using food dyes of different colors, we assess the mixing capacities of laminar flow, droplet, and bubble (Figure 5c–e). It is demonstrated that in contrast to the laminar flow, both droplets and bubbles can ensure thorough mixing within a fluidic chip. The mixing efficiency of a droplet depends significantly on the driving flow rate, and the bubbles as stirrers in a bulk solution quickly generate fully mixed solutions even at a low driving flow rate (i.e., 0.1 mm/s). Our results suggest that bubbles as an active stirrer (i.e., self-orientation) can generate strong chaotic advection, even within a straight channel, which is a key attractive feature of the proposed fluidic mixer.

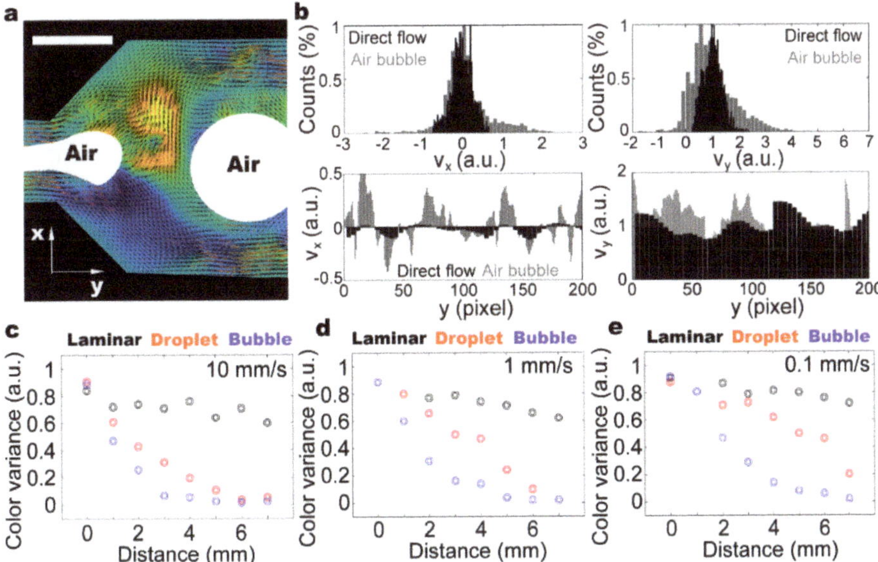

Figure 5. (**a**) Turbulent flow generated with the presence of air bubbles on-chip. (**b**) Flow distribution within the channels demonstrates that air bubbles as stirrers substantially enhance the flow rate in cross-sectional direction of the channel, which allows for fast mixing. The flow velocity is normalized to the driving flow rate to illustrate the mixing efficiency. (**c–e**) Color variances are measured in the cross-section at different positions along the fluidic channel (i.e., distance from the Y-junction). To obtain the color variances, light intensities of selected area along the fluidic channel are measured in the red channel of the RGB picture, and calculated to get the standard deviation (Figure S3). Three different driving flow rates (i.e., 10, 1, and 0.1 mm/s) are analyzed. The scale bar denotes 500 µm.

4. Conclusions

In this study, we chose not to use any external field to achieve active mixing, which allows the system to be easily fabricated and operated. Instead, chaotic advection is induced using air-bubbles as stirrers and varying channel geometry as physical obstacles. For biological and biomedical applications, in which air-bubbles can cause significant effects on the on-chip culture environment, a degassing unit can be integrated to remove the unwanted gas-liquid interface (Figure S2 and Video S3). Our proposed fluidic mixer possesses the characteristics of both active and passive mixers, allowing effective solute mixing even at a relatively low input flow rate. We believe that it will be useful for both chemical analysis and synthesis, and for studies of complex reaction networks.

Supplementary Materials: The following are available online at http://www.mdpi.com/2072-666X/11/2/195/s1. Figure S1: Fabrication process of the fluidic mixer. The template is fabricated using 3D printing. The mixed PDMS solution (base and curing agent, 10:1) is transferred into the template, which is followed by incubated at 45 °C for 24 h. The cured PDMS is then sealed with glass slides through plasma etching. Figure S2: (**a**) Optical image and (**b**) schematic showing that the air-bubbles can be effectively removed by integrating a degassing unit into the fluidic chip. (b) By collecting the mixed solution into an empty chamber, the gas and liquid can be quickly separated due to gravity. Figure S3: (**a,c**) Original optical images and (**b,d**) mono-color (red) images of the fluid channel. (**e**) Histogram of the selected area marked in Figure (**b**). (**f**) Histogram of the selected area marked in Figure (**d**). It is demonstrated that with air-bubbles as stirrers, the intensity distribution can become significantly narrower. Video S1: Movement of a droplet within a straight fluidic channel. Video S2: Chaotic advection caused by the movement of small droplets within the bulk solution. Video S3: Operation of a fluidic mixer chip integrated with a degassing unit. It is demonstrated that the air-bubbles are trapped in the large reservoir, and the "clear" mixed solution is collected elsewhere.

Author Contributions: Conceptualization, C.Z. and G.J.; methodology, C.Z.; software, G.J.; validation C.Z. and G.J.; formal analysis, X.J.; investigation, B.C.; data curation, X.J.; writing—original draft preparation, C.Z. and B.C.; writing—review and editing, C.Z. and G.J.; supervision, C.Z. and G.J.; project administration, C.Z.; funding acquisition, G.J. All authors have read and agreed to the published version of the manuscript.

Funding: This research received no external funding.

Conflicts of Interest: The authors declare no conflict of interest.

References

1. Stone, H.A.; Stroock, A.D.; Ajdari, A. Engineering flows in small devices: Microfluidics toward a lab-on-a-chip. *Annu. Rev. Fluid. Mech.* **2004**, *36*, 381–411. [CrossRef]

2. Squires, T.M.; Quake, S.R. Microfluidics: Fluid physics at the nanoliter scale. *Rev. Mod. Phys.* **2005**, *77*, 977. [CrossRef]

3. Dittrich, P.S.; Manz, A. Lab-on-a-chip: Microfluidics in drug discovery. *Nat. Rev. Drug Discov.* **2006**, *5*, 210. [CrossRef] [PubMed]

4. Duncombe, T.A.; Tentori, A.M.; Herr, A.E. Microfluidics: Reframing biological enquiry. *Nat. Rev. Mol. Cell Biol.* **2015**, *16*, 554–567. [CrossRef] [PubMed]

5. Shang, L.; Cheng, Y.; Zhao, Y. Emerging droplet microfluidics. *Chem. Rev.* **2017**, *117*, 7964–8040. [CrossRef]

6. Ahn, J.; Ko, J.; Lee, S.; Yu, J.; Kim, Y.; Jeon, N.L. Microfluidics in nanoparticle drug delivery; From synthesis to pre-clinical screening. *Adv. Drug Deliv. Rev.* **2018**, *128*, 29–53. [CrossRef]

7. Aref, H. Stirring by chaotic advection. *J. Fluid Mech.* **1984**, *143*, 1–21. [CrossRef]

8. Ottino, J.M. *The Kinematics of Mixing: Stretching, Chaos, and Transport*; Cambridge University Press: Cambridge, UK, 1989.

9. Tofteberg, T.; Skolimowski, M.; Andreassen, E.; Geschke, O. A novel passive micromixer: Lamination in a planar channel system. *Microfluid. Nanofluid.* **2010**, *8*, 209–215. [CrossRef]

10. Jones, S.W.; Thomas, O.M.; Aref, H. Chaotic advection by laminar flow in a twisted pipe. *J. Fluid Mech.* **1989**, *209*, 335–357. [CrossRef]

11. Stroock, A.D.; Dertinger, S.K.; Ajdari, A.; Mezić, I.; Stone, H.A.; Whitesides, G.M. Chaotic mixer for microchannels. *Science* **2002**, *295*, 647–651. [CrossRef]

12. Simonnet, C.; Groisman, A. Chaotic mixing in a steady flow in a microchannel. *Phys. Rev. Lett.* **2005**, *94*, 134501. [CrossRef] [PubMed]

13. Hwang, W.R.; Jun, H.S.; Kwon, T.H. Experiments on chaotic mixing in a screw channel flow. *AIChE J.* **2002**, *48*, 1621–1630. [CrossRef]

14. Alvarez-Hernández, M.M.; Shinbrot, T.; Zalc, J.; Muzzio, F.J. Practical chaotic mixing. *Chem. Eng. Sci.* **2002**, *57*, 3749–3753. [CrossRef]

15. Ansari, M.A.; Kim, K.Y.; Anwar, K.; Kim, S.M. Vortex micro T-mixer with non-aligned inputs. *Chem. Eng. J.* **2012**, *181*, 846–850. [CrossRef]

16. Li, J.; Xia, G.; Li, Y. Numerical and experimental analyses of planar asymmetric split-and-recombine micromixer with dislocation sub-channels. *J. Chem. Technol. Biotechnol.* **2013**, *88*, 1757–1765. [CrossRef]

17. Afzal, A.; Kim, K.Y. Convergent-divergent micromixer coupled with pulsatile flow. *Sens. Actuators B Chem.* **2015**, *211*, 198–205. [CrossRef]
18. Lim, T.W.; Son, Y.; Jeong, Y.J.; Yang, D.Y.; Kong, H.J.; Lee, K.S.; Kim, D.P. Three-dimensionally crossing manifold micro-mixer for fast mixing in a short channel length. *Lab Chip* **2011**, *11*, 100–103. [CrossRef]
19. Li, L.; Chen, Q.D.; Tsai, C.T. Three dimensional triangle chaotic micromixer. *Adv. Mater. Res.* **2014**, *875*, 1189–1193. [CrossRef]
20. Wu, S.J.; Hsu, H.C.; Feng, W.J. Novel design and fabrication of a geometrical obstacle-embedded micromixer with notched wall. *Jpn. J. Appl. Phys.* **2014**, *53*, 97201. [CrossRef]
21. Cubaud, T.; Ho, C.M. Transport of bubbles in square microchannels. *Phys. Fluids* **2004**, *16*, 4575–4585. [CrossRef]
22. Bottausci, F.; Mezić, I.; Meinhart, C.D.; Cardonne, C. Mixing in the shear superposition micromixer: Three-dimensional analysis. *Philos. Trans. R. Soc. Lond. Ser. A Math. Phys. Eng. Sci.* **1818**, *362*, 1001–1018. [CrossRef] [PubMed]
23. Niu, X.; Lee, Y.K. Efficient spatial-temporal chaotic mixing in microchannels. *J. Micromech. Microeng.* **2003**, *13*, 454. [CrossRef]
24. Okkels, F.; Tabeling, P. Spatiotemporal resonances in mixing of open viscous fluids. *Phys. Rev. Lett.* **2004**, *92*, 38301. [CrossRef] [PubMed]
25. Oddy, M.H.; Santiago, J.G.; Mikkelsen, J.C. Electrokinetic instability micromixing. *Anal. Chem.* **2001**, *73*, 5822–5832. [CrossRef]
26. Ho, C.M. Fluidics-the link between micro and nano sciences and technologies. In Proceedings of the 14th IEEE International Conference on Micro Electro Mechanical Systems, Interlaken, Switzerland, 25 January 2001.
27. Song, H.; Tice, J.D.; Ismagilov, R.F. A microfluidic system for controlling reaction networks in time. *Angew. Chem. Int. Ed.* **2003**, *42*, 768–772. [CrossRef]
28. Song, H.; Bringer, M.R.; Tice, J.D.; Gerdts, C.J.; Ismagilov, R.F. Experimental test of scaling of mixing by chaotic advection in droplets moving through microfluidic channels. *Appl. Phys. Lett.* **2003**, *83*, 4664–4666. [CrossRef]
29. Feng, Y.B.; Wang, B.Q.; Tian, Y.; Chen, H.; Liu, Y.G.; Fan, H.M.; Wang, K.G.; Zhang, C. Active fluidic chip produced using 3D-printing for combinatorial therapeutic screening on liver tumor spheroid. *Biosens. Bioelectron.* **2020**, *151*, 111966. [CrossRef]
30. Smolianski, A.; Haario, H.; Luukka, P. Numerical study of dynamics of single bubbles and bubble swarms. *Appl. Math. Model.* **2008**, *32*, 641–659. [CrossRef]
31. Handique, K.; Burns, M.A. Mathematical modeling of drop mixing in a slit-type microchannel. *J. Micromech. Microeng.* **2001**, *11*, 548. [CrossRef]

 *micromachines*

Article

Numerical Simulation and Experimental Validation of Liquid Metal Droplet Formation in a Co-Flowing Capillary Microfluidic Device

Qingming Hu [1,2,*], Tianyi Jiang [1] and Hongyuan Jiang [1,*]

[1]   School of Mechatronics Engineering, Harbin Institute of Technology, West Da-zhi Street 92, Harbin 150001, China; jty_hit@sina.com
[2]   School of Mechatronics Engineering, Qiqihar University, Wenhua Street 42, Qiqihar 161006, China
*   Correspondence: qminghu@gmail.com (Q.H.); jhy_hit@sina.com (H.J.); Tel.: +86-451-8641-8028 (Q.H. & H.J.)

Received: 7 January 2020; Accepted: 21 January 2020; Published: 5 February 2020

Abstract: A two-phase flow axisymmetric numerical model was proposed to understand liquid metal droplet formation in a co-flowing capillary microfluidics device based on a phase field model. The droplet detachment processes were observed in the experiment and are in good agreement with the simulation method. The effects of the viscosities and flowrates of the continuous phase fluid, interfacial tension as well as the wetting property of the metallic needle against the bulk liquid metal on the droplet formation and production rate were numerically investigated. It was found that the droplet diameter decreased with the increment of the viscosities and flowrates of the outer phase carrier fluid. The dispersed phase fluid with high interfacial tension tended to prolong the time for equilibrium between the viscous drag force and interfacial tension on the liquid–liquid fluid surface, delaying the droplet to be pinched off from the capillary orifice and causing large droplet diameter. Finally, the wetting performance of the metallic needle against the liquid metal was explored. The result indicate that the droplet diameter became less dependent on the contact angle while the size distribution of the liquid metal droplet was affected by their wetting performance. A more hydrophilic wetting performance were expected to prepare liquid metal droplet with more monodispersity. The numerical model and simulation results provide the feasibility of predicting the droplet formation with a high surface tension in a glass capillary microfluidic device.

Keywords: droplet formation; phase field model; interfacial tension; glass capillary microfluidic device

1. Introduction

With so many extraordinary physical and chemical properties, such as low melting point, low vapor pressure, high electrical/thermal conductivity and high surface tension, the liquid metals are very useful and highly potential in soft and stretchable electronics [1], heat transfer management devices [2] and functional composites [3], e.g., ultra-soft and compliant electrodes [4], interconnects [5], electrochemical sensors [6], three-dimensional printing [7,8], smart actuators and shape-memory alloys [9]. In particular, liquid metal microdroplets with a symmetrical spatial structure are useful in developing novel microfluidics engineering devices, advanced functional electronics. In addition, the liquid metal microdroplets with highly monodispersity possess tremendous application potentials in various fields of microfluics actuators, periodic structures and optics devices, for instance, microswitches [10–12], micropumps [13], micromixer [14], self-powered acceleration sensor [15], radio frequency resonators [16] and reconfigurable optical diffraction gratings [17–19]. Therefore, the fabrication of liquid metal droplets with uniform size is of great significance.

It is well known that the droplet size distribution mainly determined by the particular mechanism of droplet generation. Due to the large surface tension and high density, it is difficult to split the bulk liquid metal into microdroplets with the traditionally bulk emulsification method. Several approaches have been developed to fabricate liquid metal microspheres, such as molding [20], sonication [21] and acoustic waves [22]. The molding technique is based on the topographical mold featured by patterned cylindrical reservoirs with which the bulk liquid metal was spreading onto elastomeric sheet and flowed into the reservoirs. With the oxide skin over the droplet surface stabilize the shape firstly, the acid was employed to remove the oxide skin, enabling the totally spherical shape. Despite the simplicity of preparing the liquid metal droplets with the molding, the resulting liquid metal microdroplets size is limited to 100 μm and the molding process is not dynamical adjustable. Simultaneously, it could be time-consuming for large batch production. To improve massive production rates, the sonication technology was adopted to split bulk liquid metal into micro- to nanoscale spheres in the presence of ligands by inserting ultrasonication probe into a nonsolvent. While compared with the molding approach, the sonication method provided limited control over the droplet monodispersity and usually produce liquid metal droplets with a broad droplet size distribution, ranging from less than a few hundred nanometers to several micrometers. Additionally, the bulky sonication device makes it intractable to integrate with other detection and sensing equipment for lab on a chip application. As an alternative strategy, the acoustic wave-induced forces was explored to generate liquid metal microspheres with controllable size by adjusting the interfacial tension of the metal through the external excitation voltage in the acoustic field. The piezoelectric transducer and the signal generator devices maybe increase the cost and the system complexity. Therefore, it is extremely urgent to seek out a preferred technique to produce uniform liquid metal microspheres on demand. Microfluidic technology provides fast, high-throughput and better control over the droplet size distribution, which are useful and highly potential in various applications in many fields such as biochemical and materials science [23–25]. Typically, flow-focusing and co-flowing droplet formation mechanism-based droplet microfluidics have been developed for the preparation of uniform-sized liquid metal microspheres [26–33]. When the two neighboring immiscible fluids coming across a small orifice in the microchannel, the flow-focusing based microfluidic chip split the bulk liquid metal into microspheres with the continuous phase fluid pinching off the dispersed phase liquid metal. The droplet diameter and the generation frequency are determined by the balance between the interfacial tension and the viscous shear force exerted by the continuous phase on the dispersed phase [30]. The microsphere size can be tuned by adjusting the shearing rate of two immiscible fluids, which is determined by the flowrate, viscosity of the continuous phase and the chip geometry. The commonly used chip geometry for liquid metal preparation is X-mixer. Michael D. Dickey et al. experimentally studied the formation of galinstan liquid metal microdroplets with high viscosity continuous phase fluid pinching off the bulk liquid metal. They also investigated the influence of flowrates ratios, outer continuous phase fluid viscosities, chip geometries, and interfacial tension on the droplet size [30]. On the other hand, the electric field was implemented to rapidly vary and decrease the liquid metal microdroplets size under the same experimental conditions by changing the interfacial tension of the metal through both electrochemistry and electrocapillarity.

Although flow-focusing-based droplet technology can produce high interfacial tension liquid metal microdroplets with a relative high monodispersity, the fabrication procedures of commonly used polydimethylsiloxane (PDMS) microchannel [34] are complicated and the continuous phase organic solvents may sometimes inflate the soft PDMS microchannel, which may influence the droplet size distribution when the flowrates of the injecting flows are relatively high. As the capillary-based microfluidic device has a strong resistance to organic solvents and aggressive chemical reactions, the droplets can be prepared with less preparation costs and high accuracy by using co-flow devices, which has drawn much attention by research communities. Meanwhile, the strategy of placing wire inside the inlet microchannel was introduced to improve the droplet uniformity and provide better control over the droplet size [35–37]. To further improve the monodispersity of generated liquid metal microspheres, we therefore previously put forward a micro-needle induced strategy for the fabrication of liquid metal droplets in a co-flowing capillary microfluidic device in which a stainless steel micro-needle was inserted into the inner liquid metal phase in the glass capillary [31]. The experimental investigation showed that highly uniform liquid metal microdroplets can be obtained.

As the interfacial hydrodynamic behavior of droplet generation is affected by many parameters, mainly including the flowrates of the continuous and dispersed phase, fluid viscosities, interfacial tension, we can dynamically control the dripping mode generation and the corresponding droplet size by adjusting the aforementioned parameters. Usually, the theoretical analysis and experimental investigation was adopted to acquire the flow phenomenon in the capillary-based microfluidic device and predict the droplets sizes. Due to the complexity of theoretical analysis and the high cost of conducting experimental study, the numerical simulation has been performed to investigate such complex phenomenon in the co-flowing device. For instance, Shaowei Li et al. [38] employed a modified level set method to investigate the flow pattern transition and droplet breakup dynamics in a coaxial microchannel. Deng [39] exploited the volume-of-fluid/continuum-surface-force method to simulate the hydrodynamics of oil-in-water droplet formation in a co-flowing capillary device and systematically discussed the effect of interfacial tension, wetting properties of the capillary, and the velocities and viscosities of the two inlet fluids on the droplet size.

However, the current simulation studies mainly concentrated on the water-in-oil or oil-in-water droplet formation, and the majority of studies have focused on the experimental investigation of liquid metal microdroplets formation. The research on the numerical investigation on the liquid metal microsphere formation with microfluidic technology is very sparse, especially with a microfiber inside the inner capillary to induce the stable droplet formation, while it is also of great importance in predicting the droplet size and saving the fabrication cost. Herein, we presented the numerical investigation on the liquid metal microdroplets generation in co-flowing capillary device with a micro-needle in the inlet capillary. The effects of the interfacial tension, wetting properties of the micro-needle, and the viscosities and flowrate of the continuous phases on the droplet size were systematically studied.

2. Experiments

2.1. Materials

The glycerol (purchased from Aladdin, Shanghai, China) aqueous solution and Galinstan liquid metal (purchased from Sigma, St. Louis, MI, USA) were adopted as the continuous and dispersed phase, respectively. To avoid the spontaneous coalescence of generated liquid metal droplets, the surfactant Poly(vinyl alcohol) (PVA, 87-89 hydrolyzed, average molecular weight (MW), MW = 13,000–23,000) aqueous solution was added into the glycerol solution. The weight ratio of glycerol and 5% PVA aqueous solution was 10:2.9. Therefore, the dynamic viscosity for the two-phase fluids were defined as 0.044 Pa·s and 0.002 Pa·s at room temperature [31], respectively. The interfacial tension between the liquid metal and the glycerol aqueous solution was 0.534 N/m [28]. With a strong chemical attack resistance, a good rigidity and high tensile strength, the 1Cr18Ni19Ti stainless steel micro-needle (Zongsheng, Harbin, China) with a diameter of 70 μm was utilized as the guiding wire.

2.2. Experimental Setup

As depicted in Figure 1 (the experimental setup could be found in Figure S1), the capillary-based co-flowing microfluidic device was manufactured by assembling two tapered glass capillaries insides a square channel (AIT Glass, Inc., Largo, FL, USA, 810-9917). The two inner capillaries were concentric aligned with axial spacing of 100 μm, and the metallic needle was inserted into the inlet capillary along the axis of the orifice, with a transverse distance between the micro-needle tip and the center of the capillary orifice of 175 μm. The detailed dimensions and fabrication processes of the micro-needle induced co-flowing glass capillary microfluidic device could be obtained in our previous published paper [31]. The glycerol and liquid metal were supplies separately by two microsyringe pumps (Harvard Apparatus, Holliston, MA, USA) equipped with gastight precision glass syringes (Hamilton,

Bonaduz, Switzerland). An optical microscope (CKX41, Olympus, Tokyo, Japan) with a high-speed charge coupled device (CCD) video camera (DP27, Olympus, Japan) connected was utilized to record the flow pattern and the droplet formation in the microdevice.

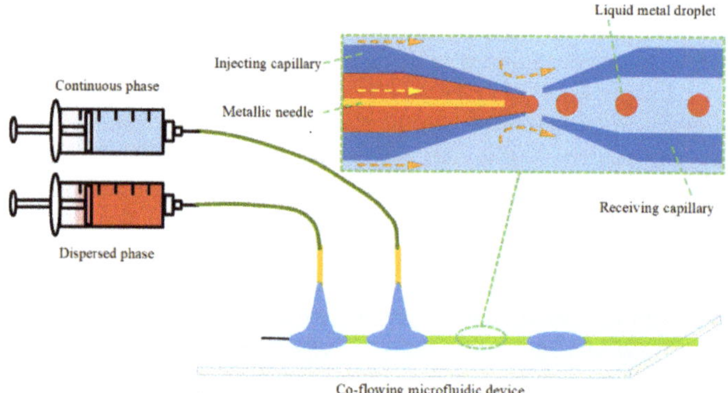

Figure 1. Schematic of the micro-needle induced co-flowing microfluidic experimental setup.

3. Numerical Simulation

3.1. Governing Equations

A two-dimensional axisymmetric numerical model was developed to simulate the liquid droplet emulsion formation. The transient flows of the continuous and dispersed phases in the co-flowing capillary are governed by the incompressible Navier–Stokes equation and continuity equation [40].

$$\rho\frac{\partial \mathbf{u}}{\partial t} + \rho(\mathbf{u}\cdot\nabla\mathbf{u}) = \nabla\cdot\left[-p\mathbf{I} + \mu\nabla\left(\nabla\mathbf{u} + \nabla\mathbf{u}^{T}\right)\right] + F \tag{1}$$

$$\nabla\mathbf{u} = 0 \tag{2}$$

where ρ and $\mathbf{u}$ are the density and velocity vector, respectively, while t, p and I are the time, pressure and identity matrix. μ denotes the dynamics viscosity of the fluid. F signifies body force/source force, mainly includes gravitational force, inertial force and interfacial force. While the droplet diameter is in the order of micro-scale in the co-flowing capillary, the gravitational force and the inertial force can be neglected compared with the interfacial force, The interfacial tension force acting on the liquid–liquid interface between two immiscible fluids can be expressed as follow [41,42]:

$$F = \sigma\kappa\delta n \tag{3}$$

where σ is the surface tension coefficient, k and δ denote the curvature of the interface and the function concentrated across the neighboring immiscible fluids interface, respectively. n is the unit interface normal vector pointing into the droplet, which can be obtained through the flowing equation [42]:

$$n = \frac{\nabla\phi}{|\nabla\phi|} \tag{4}$$

where ϕ is the phase field function, which is applied to describe the fluid–fluid interface between two neighboring immiscible mediums and can be calculated by the following advection equations [43].

$$\frac{\partial \phi}{\partial t} + \mathbf{u}\cdot\nabla\phi = \nabla\cdot\frac{\gamma\lambda}{\varepsilon^2}\nabla\psi \tag{5}$$

$$\psi = -\nabla \cdot \varepsilon^2 \nabla \phi + \left(\phi^2 - 1\right)\phi \tag{6}$$

where γ, λ and ε are the numerical stabilization parameters, which are adopted to control the interface thickness, mobility tuning and define reinitialization parameter, respectively. To minimize the numerical oscillations, the phase field function ϕ is utilized to smooth the fluid properties across the two neighboring immiscible fluids and track the profile of the liquid–liquid interface [44].

$$\rho = \rho_{continuous} + \left(\rho_{dispersed} - \rho_{continuous}\right)\phi \tag{7}$$

$$\mu = \mu_{continuous} + \left(\mu_{dispersed} - \mu_{continuous}\right)\phi \tag{8}$$

The density ρ and dynamics viscosity μ in Equation (1) can be obtained with the above Equations (7) the (8). The phase field function denotes fractional volume of the continuous and dispersed phases in a computational cell. Herein, $\phi = 1$ and $\phi = 0$ represent the cell is filled with dispersed phase and continuous phase, respectively, while $0 < \phi < 1$ indicates the fluid–fluid interface existing in the cell.

3.2. Numerical Method

We conducted the numerical simulation in order to explore the flow pattern and the liquid metal microsphere formation simulation using a commercial software package (COMSOL Inc., Stockholm, Sweden). As depicted in Figure 2, we created an axisymmetric simulation geometry of the flow domain for the three-dimensional co-flowing capillary. The glycerol aqueous solution was introduced into the annular clearance between the outer square capillary and inner circular capillary, while the Galinstan bulk liquid metal was injected into the inner round capillary.

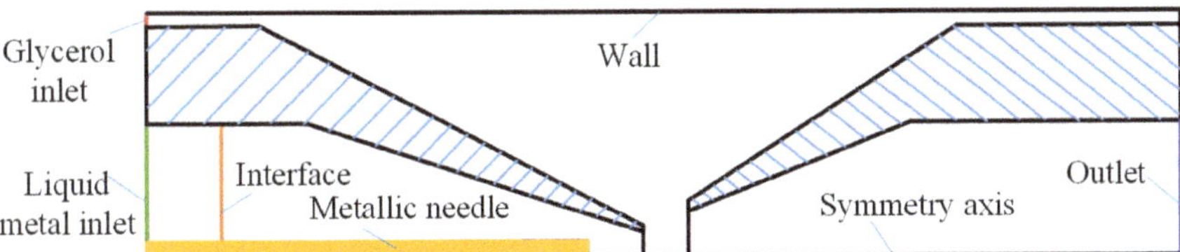

Figure 2. Geometry and boundary condition of liquid metal microsphere formation simulation with the micro-needle induced strategy in co-flowing microchannel.

The influence of wetted wall on the droplet formation can defined by contact angle between the two neighboring fluids in the simulation. Arising from the effect of metallic affinity, we determined the wetting property of inner phase bulk Galinstan liquid metal on the micro-needle by defining the static contact angle between the liquid metal and the micro-needle, which is shown in Figure 3 as α_1. On the other side, the contact angle between the liquid metal and the capillary wall was described as α_2. The initial contact angle between the continuous phase and the micro-needle and the glass wall were determined as pi/4 and 5 pi/18, respectively [45].

Due to the small Reynolds number in the co-flowing capillary microchannel, the laminar flow model was utilized. The phase field function was set as an additional variable in the simulation and adopted to track the two-phase immiscible fluids interface structure. The advection term was used to calculate the momentum and the phase field function. The mobility tunning λ parameter was set as 1. The inlets of continuous phase and dispersed phase were defined by the normal inflow speed calculated from volume flowrate, while the outlet was exerted as the opening boundaries with one atmospheric pressure. The capillary microchannel sidewalls and the metallic needle surface were imposed as the non-slip condition. Since the grid may significantly influence the droplet formation simulation result, a local refinement grid was adopted around the contact surfaces between the continuous phase and the inner capillary wall and the micro-needle surface to capture the free surface–surface change more accurately and smoothly. The elements number was 58,044 in the computational domain. In the following calculation, we found that the results variation caused with the mesh refinement is no more than 4%.

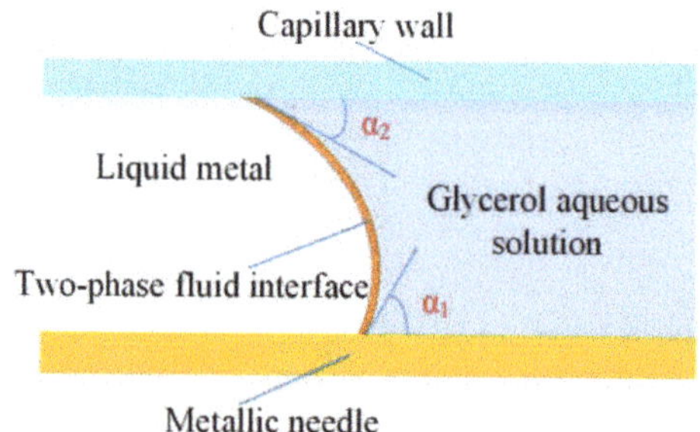

Figure 3. The contact angle between the liquid metal and the metallic needle, and capillary wall, respectively.

4. Results and Discussion

4.1. Model Validation of Simulation Results

A series of numerical simulations and experiments were conducted to observe the droplet morphology variations during the liquid metal droplet formation. We defined the initial interfacial tension $\sigma = 0.534$ N/m, the viscosity of the continuous phase and the disperse phase were 0.044 Pa·s and 0.002 Pa·s, respectively. Figure 4 presents the simulated and experimental snapshot of droplet formation process in the coaxial microchannel. On the whole, the simulated two-phase fluids interface movement was in accordance with the experiment observation results, which validated that the phase-field method was capable of predicting the morphological variations during the liquid metal microdroplet formation. Due to the large interfacial tension and high viscosity of the Galinstan liquid metal, the droplet initially attached to the capillary tip and kept growing with the bulk liquid metal continuously being injected, which, in some cases, accentuated the effect of contact line dynamics. The necking of the bulk liquid metal was initiated once the glycerol aqueous solutions was intruded near the inlet capillary orifice. Simultaneously, the neck became thin with the increment of the viscous drag force. The liquid metal droplet detached from the capillary orifice when the viscous drag force was large enough to oppose the action of the surface tension. As the Galinstan liquid metal had a large surface tension, the produced droplet size was relatively bigger than the traditional oil in water or water in oil droplet module under the same circumstances [28].

To further validate the effectiveness of the proposed phase-field method-based two-phase flow axisymmetric numerical model, droplet formation simulations were conducted for different flowrates of continuous phase when the flowrate of dispersed phase was defined as 10 μL/min, ranging from 0.070 to 0.100 mL/min. The droplet diameters in the experiment were compared with the numerical results. It was predicted from Figure 5 that droplet size decreased as the flowrate of the continuous phase increased. The variation tendency between the droplet size and the flowrate of the continuous phase was also in agreement with the experimental results published by Hutter [29]. The droplet diameter generated with the simulation was 312.3 μm when the continuous phase flowrate and the dispersed phase flowrate were 100 μL/min and 10 μL/min, respectively, which was almost consistent with the experimental result of 318 μm under the same circumstances. Additionally, the droplet sizes obtained from the experiment were slightly larger than the simulation results, whereas the discrepancy decreased with the increasing flowrate of the glycerol aqueous solution. This could be attributed to the velocity variation fluctuation of the injecting bulk liquid metal stream during the experiment. Both simulation and experiment indicated that the generated liquid metal microspheres were highly monodispersed.

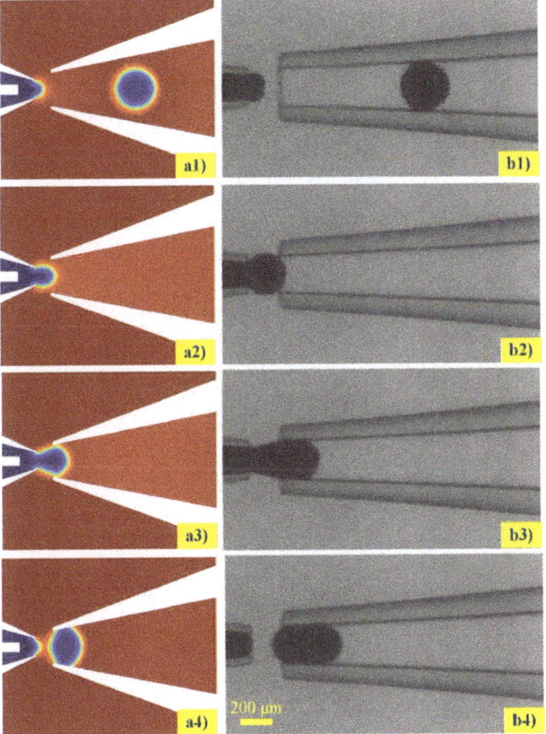

Figure 4. Comparison of liquid metal microdroplets detachment form the capillary tip between simulated (**a1–a4**) and experimental (**b1–b4**) observation during the droplet formations.

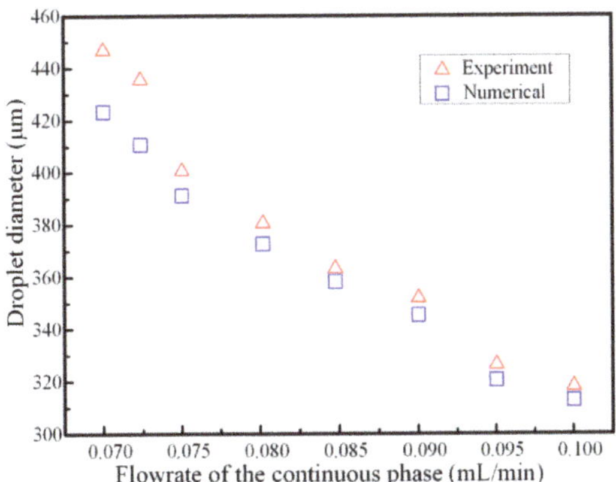

Figure 5. Comparison of the droplet diameter between numerical simulations and experimental results for different flowrates of continuous phase when the flowrate of dispersed phase was fixed as 10 μL/min.

4.2. Effect of Interfacial Tension

The interfacial tension, which was the dominating force at the microscale, was of vital role in the process of emulsification. When the two immiscible fluids intimately contacted with each other, the liquids tended to be spherical to maintain the smallest energy on the two-phase fluid interface [46]. The breakup mechanism and stability of the droplets were affected by two neighboring fluids interfacial tension in co-flowing capillary microfluidic device. Especially for the liquid metal with high surface tension, the interfacial tension could keep the liquid metal microsphere attached at the capillary tip to avoid being detached from the incoming dispersed phase stream. Figure 6 depicts the relationship between the interfacial tension on the microdroplet formation for the fixed flowrate of continuous phase and dispersed phases being as 100 µL/min and 10 µL/min, respectively. Compared with the traditional oil phase droplet formation, the simulation resulted show that the liquid metal droplet diameter was much larger and the production rate was lower increased with the increment of interfacial tension under the same circumstance. As the interfacial tension could always keep the droplet attaching to the capillary tip, the droplet volume became larger and larger with the incoming continuous phase bulk liquid metal. Then, the droplet was detached from the orifice when the equilibrium between the interfacial tension and the viscous drag force was broken. On the other hand, the size of produced microdroplets varied inversely with the dimensionless capillary number, $Ca = V\mu/\gamma$ (where γ was the surface tension at the interface between the bulk liquid metal and the carrier fluid, V and μ were the characteristic velocity and dynamic viscosity of the glycerol aqueous solution, respectively). Namely, changing the interfacial tension at the two-phase surface could allow us to adjust the liquid metal microspheres diameter. Therefore, it could be accountable that the time for reaching the equilibrium state was much longer with the increasing liquid metal surface tension, resulting in a longer droplet length and a larger diameter.

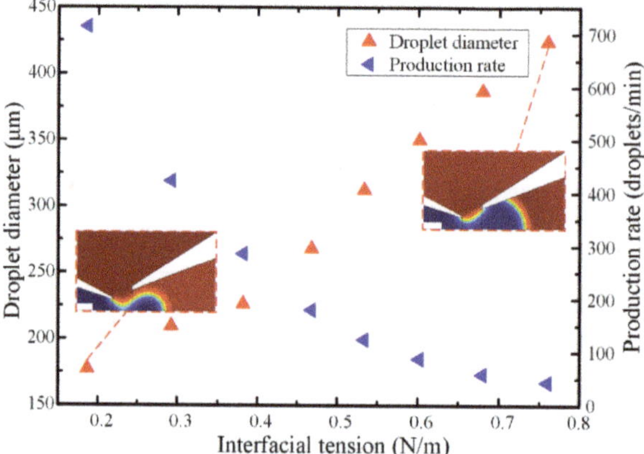

Figure 6. The effect of interfacial tension on the droplet diameter and generation frequency when the flowrate of continuous and dispersed phases were 100 µL/min and 10 µL/min, respectively.

4.3. Effect of Viscosity of Continuous Phase Fluid

The incoming fluid viscosity could affect the flow pattern and dynamic behavior of droplet formation. When the viscosity discrepancy exceeded a certain threshold, the equilibrium state between the interfacial tension and the drag force was broken. The droplet could be prepared smoothly when there were matched viscosities between the continuous phase and the disperse phase [47]. The viscosity effect on droplet formation in a micro-needle induced co-flowing microfluidic device was investigated numerically. At a lower continuous fluid viscosity, which was about 10 times the inner phase, the fluid pair behaved more like a single fluid and the shearing force acting on the bulk liquid metal were much lower, causing a long time for the droplets to be pinched off from the capillary tip and hence produced a larger droplet diameter. By increasing the viscosity of continuous fluid, the viscous shear stress acting on the boundary of neighboring immiscible fluids increased, causing droplets formation to occur much quickly. The greater discrepancy between the neighboring fluids viscosities, the bigger viscous shear acting on the fluid interface, resulting in a shorter detachment time and much smaller droplets. The results (shown in Figure 7) indicate that the liquid metal droplet diameter was decreased from 423.2 to 245.3 µm with the increase of viscosity ratios from 10 to 55 under the same flow condition $Q_w = 100$ µL/min, $Q_o = 10$ µL/min. The effect of increasing the continuous phase viscosity also increased the rate of droplet production. With the decreasing in liquid metal droplet volumes as the outer phase fluid viscosity increased, the droplet generation frequency increased to account for the decreased droplet volume. Additionally, the bulk liquid metal was sheared off more quickly due to the larger continuous fluid viscosity, leading to an increase in the droplet frequency.

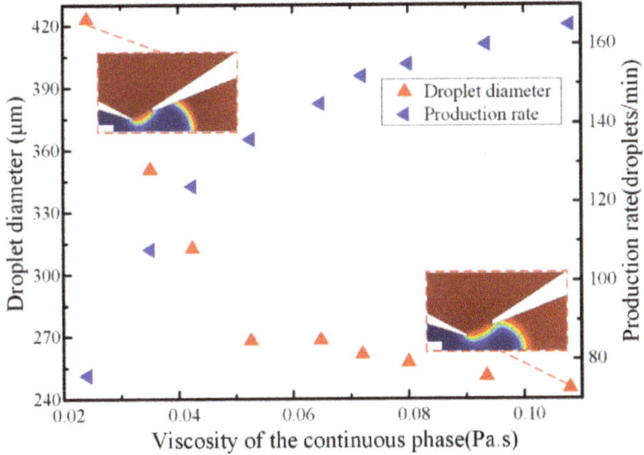

Figure 7. The effect of the viscosity of continuous phase on the droplet diameter and generation frequency when the flowrate of continuous and dispersed phases were 100 µL/min and 10 µL/min, respectively.

4.4. Effect of the Flowrates of Continuous Phase

The droplet size could be controlled by adjusting the flowrate of the inlet immiscible flows. The lower continuous fluid flowrate led to a larger droplet size and a lower generation frequency. Figure 8 depicts the effect of continuous phase flowrates on the droplet formation with the dispersed phase flowrate fixed at 10 µL/min and the viscosity of continuous phase were set as 10 µL/min and 0.044 Pa·s, respectively. The droplet diameter decreased with the increase of the continuous phase flowrate, which was in good accordance with the relationship between the capillary number and the droplet size. Simultaneously, it could be observed from the simulation snapshot that the droplet length became shorter as the continuous phase flowrate increased. This could be ascribed to the higher velocity gradient generated on the liquid metal–glycerol interface under the higher continuous fluid velocity, which caused a higher viscous drag force acting on the droplet and hence, smaller droplets could be obtained. When the flowrate of the glycerol aqueous solution was less than 0.085 mL/min at the fixed dispersed phase flowrate of 10 µL/min, the droplet production augmented significantly linearly proportionally to the flowrate. That was because production frequency and the droplet volume must be inversely related to maintain the flow continuity in the capillary microchannel under the low outer phase fluid flowrate, which was also in good accordance with the previously published conclusion [48]. However, at the fixed dispersed phase flowrate, the bulk liquid metal stream was suppressed and could not smoothly flow out of the inner capillary when the flowrate of the continuous phase was large. The competition between the drag force and the surface tension on the liquid metal–glycerol aqueous solution interface was broken with the flowrate of the continuous phase above a certain critical value.

The liquid metal microsphere was detached from the capillary tip and then the liquid metal stream was retracted back into the inner capillary due to the high continuous fluid velocity and flow instability. Therefore, the hydrodynamic instability of the liquid metal stream is significant during the necking and pinch-off stage when the flowrate of the continuous phase is high, leading to longer droplet formation time interval and hence, a lower droplets generation frequency. The simulation results of the relation between the droplet production rates and the flowrate are also in agreement with the experimental results of the previous findings [29].

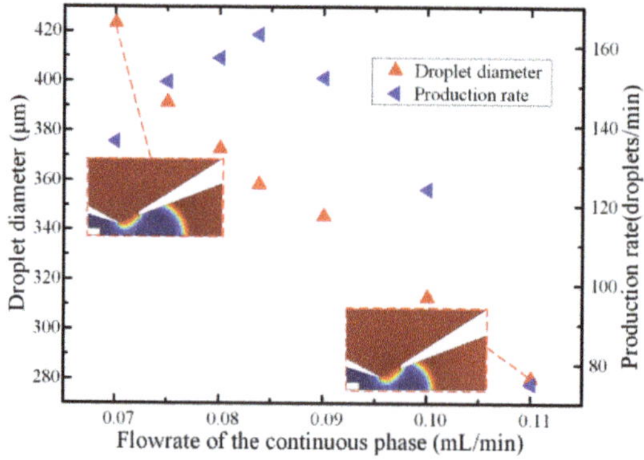

Figure 8. The effect of the flowrate of the continuous phase on the droplet diameter and generation frequency when the flowrate of dispersed phase and the viscosity of continuous phase were set as 10 µL/min and 0.044 Pa·s, respectively.

4.5. Effect of the Wetting Property of the Micro-Needle

As the micro-needle was inserted into the inner capillary, the wetting properties might influence the droplet diameter and the size distribution. We previously defined the contact angle among the metallic needle, bulk liquid metal and the glycerol solution, and herein, numerically investigated the droplet diameter with respect to the contact angle when the viscosity of continuous phase was 0.044 Pa·s. Figure 9 depicts the effect of the contact angle α_1 on the droplet diameter with a fixed contact angle α_2. On the whole, the droplet diameter declined from 314.8 µm to 312.2 µm with the contact angle increased from 50° to 120° when the flowrates of the continuous phase and the dispersed phase were fixed at 100 µL/min and 10 µL/min, respectively. Namely, the contact angle had a slight influence on the droplet size. However, after the statistical analysis of the simulated generated droplet, the variance of the droplet diameter normal distribution for the contact angle of 120° was 2.26, which is much smaller than 8.35 with a contact angle of 50°. That means the produced droplet at larger contact angle has better monodispersity compared with the smaller one. It could be worth noting that the micro-needle inside the capillary had a better wettability performance with a larger contact angle, which is good for droplet movement and adsorption on the microfiber [37,49,50]. Therefore, the flow stability of droplet formation was affected at a lower contact angle and a droplet diameter with a better distribution could be obtained with a relatively large contact angle.

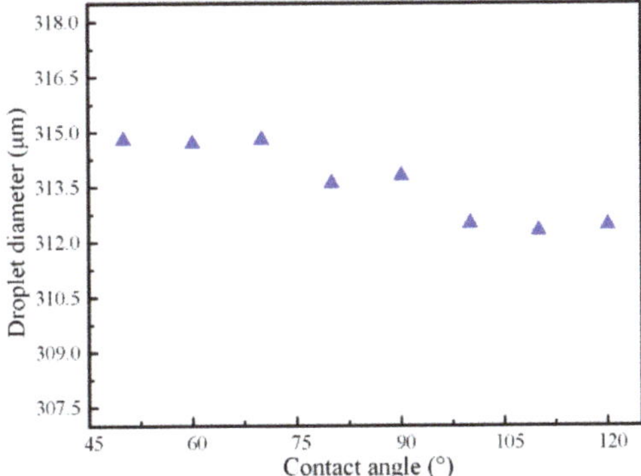

Figure 9. The effect of contact angle on the droplet diameter when the flowrates of the continuous phase and the dispersed phase were fixed at 100 μL/min and 10 μL/min, respectively.

5. Conclusions

In summary, numerical simulations of the liquid metal microspheres formation in a micro-needle induced co-flowing capillary microfluidic device were conducted using the phase-field method. The reasonableness of the computation model was experimentally validated. The relationship between the droplet size and generation frequency with respect to some important factors were numerically investigated, mainly including the interfacial tension, fluid viscosities, wetting property of the micro-needle against the bulk liquid metal and flowrates. The results demonstrate that the breakup mechanism of the liquid metal with a high surface tension was strongly dependent on the competition between the viscous drag force induced by fluid flow of the continuous phase and the interfacial tension on the two-phase fluid interface. On the whole, with the increasing viscosity and flowrate of the continuous carrier flow, a higher viscous drag force was generated and hence, the detachment time for the droplet formation was diminished, leading to the fabrication of smaller liquid metal droplets with a high generation frequency. The smaller capillary number determined by the higher interfacial tension delayed the balance between the drag force and the interfacial tension for the liquid metal droplet to be pinched off from the capillary tip. Accordingly, larger liquid metal droplets were prepared with more inner phase fluid flowing into the droplet under high surface tension. Although the contact angle of micro-needle against liquid metal had a slight influence on the droplet diameter, the size distribution was affected by the contact angle. A droplet of a more uniform size could be obtained when the contact angle was larger. That was probably attributed to the better wettability between the round metallic needle and the bulk liquid metal. The presented axisymmetric numerical model and the simulation results above could provide a reliable prediction for preparation of droplet in a controllable way in co-flowing microfluidic devices.

Supplementary Materials: The following are available online at https://www.mdpi.com/2072-666X/11/2/169/s1, Figure S1: The image of the experimental setup.

Author Contributions: Q.H. designed and analyzed droplet breakup mechanism and wrote the paper, T.J. conducted the simulation and data analysis, and H.J. provided the ideas and edited the draft of the paper. All authors have read and agreed to the published version of the manuscript.

Funding: This work is supported by the National Natural Science Foundation of China (Grant No. 11872165), the University Nursing Program for Young Scholars with Creative Talents in Heilongjiang Province (Grant No. UNPYSCT-2018104), the Industry-Academic Cooperative Education Project of Ministry of Education (Grant No. 201901145012), and Research Project on Higher Education Reform in Heilongjiang Province (Grant No. SJGY20190715).

Conflicts of Interest: The authors declare no conflict of interest.

References

1. Dickey, M.D. Stretchable and Soft Electronics using Liquid Metals. *Adv. Mater.* **2017**, *29*, 1606425.
2. Deng, Y.; Liu, J. A liquid metal cooling system for the thermal management of high power LEDs. *Int. Commun. Heat Mass Transf.* **2010**, *37*, 788–791.
3. Chen, Y.; Zhou, T.; Li, Y.; Zhu, L.; Handschuh-Wang, S.; Zhu, D.; Zhou, X.; Liu, Z.; Gan, T.; Zhou, X. Robust Fabrication of Nonstick, Noncorrosive, Conductive Graphene-Coated Liquid Metal Droplets for Droplet-Based, Floating Electrodes. *Adv. Funct. Mater.* **2018**, *28*, 1706277.
4. Shay, T.; Velev, O.D.; Dickey, M.D. Soft electrodes combining hydrogel and liquid metal. *Soft Matter.* **2018**, *14*, 3296–3303.
5. Suarez, F.; Parekh, D.P.; Ladd, C.; Vashaee, D.; Dickey, M.D.; Öztürk, M.C. Flexible thermoelectric generator using bulk legs and liquid metal interconnects for wearable electronics. *Appl. Energy* **2017**, *202*, 736–745.
6. Sheng, L.; Teo, S.; Liu, J. Liquid-Metal-Painted Stretchable Capacitor Sensors for Wearable Healthcare Electronics. *J. Med. Biol. Eng.* **2016**, *36*, 265–272.
7. Wang, X.; Zhang, Y.; Guo, R.; Wang, H.; Yuan, B.; Liu, J. Conformable liquid metal printed epidermal electronics for smart physiological monitoring and simulation treatment. *J. Micromech. Microeng.* **2018**, *28*, 034003.
8. Yu, Y.-Z.; Lu, J.-R.; Liu, J. 3D printing for functional electronics by injection and package of liquid metals into channels of mechanical structures. *Mater. Des.* **2017**, *122*, 80–89.
9. Mailen, R.W.; Dickey, M.D.; Genzer, J.; Zikry, M. Effects of thermo-mechanical behavior and hinge geometry on folding response of shape memory polymer sheets. *J. Appl. Phys.* **2017**, *122*, 195103.
10. Wissman, J.; Dickey, M.D.; Majidi, C. Field-Controlled Electrical Switch with Liquid Metal. *Adv. Sci.* **2017**, *4*, 1700169.
11. Baek, S.; Park, U.; Choi, I.H.; Kim, J. Pneumatic RF MEMS switch using a liquid metal droplet. *J. Micromech. Microeng.* **2013**, *23*, 055006.
12. McLanahan, A.R.; Richards, C.D.; Richards, R.F. A dielectric liquid contact thermal switch with electrowetting actuation. *J. Micromech. Microeng.* **2011**, *21*, 104009.
13. Tang, S.-Y.; Khoshmanesh, K.; Sivan, V.; Petersen, P.; O'Mullane, A.P.; Abbott, D.; Mitchell, A.; Kalantar-zadeh, K. Liquid metal enabled pump. *Proc. Natl. Acad. Sci. USA* **2014**, *111*, 3304–3309. [PubMed]
14. Tang, S.-Y.; Sivan, V.; Petersen, P.; Zhang, W.; Morrison, P.D.; Kalantar-zadeh, K.; Mitchell, A.; Khoshmanesh, K. Liquid Metal Actuator for Inducing Chaotic Advection. *Adv. Funct. Mater.* **2014**, *24*, 5851–5858. [CrossRef]
15. Zhang, B.; Zhang, L.; Deng, W.; Jin, L.; Chun, F.; Pan, H.; Gu, B.; Zhang, H.; Lv, Z.; Yang, W.; et al. Self-Powered Acceleration Sensor Based on Liquid Metal Triboelectric Nanogenerator for Vibration Monitoring. *ACS Nano* **2017**, *11*, 7440–7446. [PubMed]
16. Zou, L.; Withayachumnankul, W.; Shah, C.M.; Mitchell, A.; Bhaskaran, M.; Sriram, S.; Fumeaux, C. Dielectric resonator nanoantennas at visible frequencies. *Opt. Express* **2013**, *21*, 1344–1352.
17. Hashimoto, M.; Mayers, B.; Garstecki, P.; Whitesides, G.M. Flowing Lattices of Bubbles as Tunable, Self-Assembled Diffraction Gratings. *Small* **2006**, *2*, 1292–1298.
18. Yu, J.; Yang, Y.; Liu, A.; Chin, L.; Zhang, X. Microfluidic droplet grating for reconfigurable optical diffraction. *Opt. Lett.* **2010**, *35*, 1890–1892.
19. Pollack, M.G.; Shenderov, A.D.; Fair, R.B. Electrowetting-based actuation of droplets for integrated microfluidics. *Lab Chip* **2002**, *2*, 96–101.
20. Mohammed, M.; Xenakis, A.; Dickey, M. Production of Liquid Metal Spheres by Molding. *Metals* **2014**, *4*, 465–476.
21. áO'Mullane, A.P. Generation of catalytically active materials from a liquid metal precursor. *Chem. Commun.* **2015**, *51*, 14026–14029.
22. Tang, S.Y.; Ayan, B.; Nama, N.; Bian, Y.; Lata, J.P.; Guo, X.; Huang, T.J. On-Chip Production of Size-Controllable Liquid Metal Microdroplets Using Acoustic Waves. *Small* **2016**, *12*, 3861–3869. [PubMed]
23. Zhu, P.; Wang, L. Passive and active droplet generation with microfluidics: A review. *Lab Chip* **2017**, *17*, 34–75.
24. Adamo, A.; Sharei, A.; Adamo, L.; Lee, B.; Mao, S.; Jensen, K.F. Microfluidics-based assessment of cell deformability. *Anal. Chem.* **2012**, *84*, 6438–6443.

25. Tan, Y.-C.; Cristini, V.; Lee, A.P. Monodispersed microfluidic droplet generation by shear focusing microfluidic device. *Sens. Actuators B Chem.* **2006**, *114*, 350–356.

26. Tang, S.Y.; Joshipura, I.D.; Lin, Y.; Kalantar-Zadeh, K.; Mitchell, A.; Khoshmanesh, K.; Dickey, M.D. Liquid-Metal Microdroplets Formed Dynamically with Electrical Control of Size and Rate. *Adv. Mater.* **2016**, *28*, 604–609.

27. Gol, B.; Kurdzinski, M.E.; Tovar-Lopez, F.J.; Petersen, P.; Mitchell, A.; Khoshmanesh, K. Hydrodynamic directional control of liquid metal droplets within a microfluidic flow focusing system. *Appl. Phys. Lett.* **2016**, *108*, 164101.

28. Gol, B.; Tovar-Lopez, F.J.; Kurdzinski, M.E.; Tang, S.Y.; Petersen, P.; Mitchell, A.; Khoshmanesh, K. Continuous transfer of liquid metal droplets across a fluid-fluid interface within an integrated microfluidic chip. *Lab Chip* **2015**, *15*, 2476–2485.

29. Hutter, T.; Bauer, W.-A.C.; Elliott, S.R.; Huck, W.T.S. Formation of Spherical and Non-Spherical Eutectic Gallium-Indium Liquid-Metal Microdroplets in Microfluidic Channels at Room Temperature. *Adv. Funct. Mater.* **2012**, *22*, 2624–2631.

30. Thelen, J.; Dickey, M.D.; Ward, T. A study of the production and reversible stability of EGaIn liquid metal microspheres using flow focusing. *Lab Chip* **2012**, *12*, 3961–3967.

31. Hu, Q.; Ren, Y.; Zheng, X.; Hou, L.; Jiang, T.; Liu, W.; Tao, Y.; Jiang, H. A micro-needle induced strategy for preparation of monodisperse liquid metal droplets in glass capillary microfluidics. *Microfluid. Nanofluid.* **2019**, *23*, 13.

32. Lin, P.; Wei, Z.; Yan, Q.; Xie, J.; Fan, Y.; Wu, M.; Chen, Y.; Cheng, Z. Capillary-Based Microfluidic Fabrication of Liquid Metal Microspheres toward Functional Microelectrodes and Photothermal Medium. *ACS Appl. Mater. Interfaces* **2019**, *11*, 25295–25305. [PubMed]

33. Zhang, K.; Ren, Y.; Tao, Y.; Deng, X.; Liu, W.; Jiang, T.; Jiang, H. Efficient particle and droplet manipulation utilizing the combined thermal buoyancy convection and temperature-enhanced rotating induced-charge electroosmotic flow. *Anal. Chim. Acta* **2020**, *1096*, 108–119. [PubMed]

34. Lang, Q.; Wu, Y.; Ren, Y.; Tao, Y.; Lei, L.; Jiang, H. AC Electrothermal Circulatory Pumping Chip for Cell Culture. *ACS Appl. Mater. Interfaces* **2015**, *7*, 26792–26801.

35. Bangjang, T.; Cherkasov, N.; Denissenko, P.; Jaree, A.; Rebrov, E.V. Enhanced Droplet Size Control in Liquid-Liquid Emulsions Obtained in a Wire-Guided X-Mixer. *Chem. Eng. Technol.* **2019**, *42*, 1053–1058.

36. Si, T.; Feng, H.; Luo, X.; Xu, R.X. Formation of steady compound cone-jet modes and multilayered droplets in a tri-axial capillary flow focusing device. *Microfluid. Nanofluid.* **2014**, *18*, 967–977.

37. Fang, J.; Davoudi, M.; Chase, G. Drop movement along a fiber axis due to pressure driven air flow in a thin slit. *Sep. Purif. Technol.* **2015**, *140*, 77–83.

38. Lan, W.; Li, S.; Luo, G. Numerical and experimental investigation of dripping and jetting flow in a coaxial micro-channel. *Chem. Eng. Sci.* **2015**, *134*, 76–85.

39. Deng, C.; Wang, H.; Huang, W.; Cheng, S. Numerical and experimental study of oil-in-water (O/W) droplet formation in a co-flowing capillary device. *Colloids Surf. A Physicochem. Eng. Asp.* **2017**, *533*, 1–8. [CrossRef]

40. Nabavi, S.A.; Vladisavljević, G.T.; Gu, S.; Ekanem, E.E. Double emulsion production in glass capillary microfluidic device: Parametric investigation of droplet generation behaviour. *Chem. Eng. Sci.* **2015**, *130*, 183–196. [CrossRef]

41. Nabavi, S.A.; Gu, S.; Vladisavljevic, G.T.; Ekanem, E.E. Dynamics of double emulsion break-up in three phase glass capillary microfluidic devices. *J. Colloid Interface Sci.* **2015**, *450*, 279–287. [CrossRef] [PubMed]

42. Soh, G.Y.; Yeoh, G.H.; Timchenko, V. Numerical investigation on the velocity fields during droplet formation in a microfluidic T-junction. *Chem. Eng. Sci.* **2016**, *139*, 99–108. [CrossRef]

43. De Menech, M. Modeling of droplet breakup in a microfluidic T-shaped junction with a phase-field model. *Phys. Rev. E* **2006**, *73*, 031505. [CrossRef]

44. Nabavi, S.A.; Vladisavljevic, G.T.; Bandulasena, M.V.; Arjmandi-Tash, O.; Manovic, V. Prediction and control of drop formation modes in microfluidic generation of double emulsions by single-step emulsification. *J. Colloid Interface Sci.* **2017**, *505*, 315–324. [CrossRef] [PubMed]

45. Smithwich, R.W. Contact-angle studies of microscopic mercury droplets on glass. *J. Colloid Interface Sci.* **1988**, *123*, 482–485. [CrossRef]

46. Peng, L.; Yang, M.; Guo, S.S.; Liu, W.; Zhao, X.Z. The effect of interfacial tension on droplet formation in flow-focusing microfluidic device. *Biomed. Microdevices* **2011**, *13*, 559–564. [CrossRef]

47. Tice, J.D.; Lyon, A.D.; Ismagilov, R.F. Effects of viscosity on droplet formation and mixing in microfluidic channels. *Anal. Chim. Acta* **2004**, *507*, 73–77. [CrossRef]

48. Gómez-Pastora, J.; Amiri Roodan, V.; Karampelas, I.H.; Alorabi, A.Q.; Tarn, M.D.; Iles, A.; Bringas, E.; Paunov, V.N.; Pamme, N.; Furlani, E.P.; et al. Two-Step Numerical Approach to Predict Ferrofluid Droplet Generation and Manipulation inside Multilaminar Flow Chambers. *J. Phys. Chem. C* **2019**, *123*, 10065–10080. [CrossRef]

49. Manzo, G.M.; Wu, Y.; Chase, G.G.; Goux, A. Comparison of nonwoven glass and stainless steel microfiber media in aerosol coalescence filtration. *Sep. Purif. Technol.* **2016**, *162*, 14–19. [CrossRef]

50. Jamali, M.; Tafreshi, H.V.; Pourdeyhimi, B. Droplet Mobility on Hydrophobic Fibrous Coatings Comprising Orthogonal Fibers. *Langmuir* **2018**, *34*, 12488–12499. [CrossRef]

 *micromachines*

Review

Droplet Microfluidics-Enabled High-Throughput Screening for Protein Engineering

Lindong Weng * and James E. Spoonamore

enEvolv, Inc., Medford, MA 02155, USA; j.spoonamore@enevolv.com
* Correspondence: l.weng@enevolv.com

Received: 5 October 2019; Accepted: 26 October 2019; Published: 29 October 2019

Abstract: Protein engineering—the process of developing useful or valuable proteins—has successfully created a wide range of proteins tailored to specific agricultural, industrial, and biomedical applications. Protein engineering may rely on rational techniques informed by structural models, phylogenic information, or computational methods or it may rely upon random techniques such as chemical mutation, DNA shuffling, error prone polymerase chain reaction (PCR), etc. The increasing capabilities of rational protein design coupled to the rapid production of large variant libraries have seriously challenged the capacity of traditional screening and selection techniques. Similarly, random approaches based on directed evolution, which relies on the Darwinian principles of mutation and selection to steer proteins toward desired traits, also requires the screening of very large libraries of mutants to be truly effective. For either rational or random approaches, the highest possible screening throughput facilitates efficient protein engineering strategies. In the last decade, high-throughput screening (HTS) for protein engineering has been leveraging the emerging technologies of droplet microfluidics. Droplet microfluidics, featuring controlled formation and manipulation of nano- to femtoliter droplets of one fluid phase in another, has presented a new paradigm for screening, providing increased throughput, reduced reagent volume, and scalability. We review here the recent droplet microfluidics-based HTS systems developed for protein engineering, particularly directed evolution. The current review can also serve as a tutorial guide for protein engineers and molecular biologists who need a droplet microfluidics-based HTS system for their specific applications but may not have prior knowledge about microfluidics. In the end, several challenges and opportunities are identified to motivate the continued innovation of microfluidics with implications for protein engineering.

Keywords: FADS; emulsification; droplet coalescence; enzyme engineering; synthetic biology

---

1. Introduction

Engineered proteins with tailored properties evolved from natural precursors have been playing an increasingly important role in a spectrum of agricultural, industrial, and biomedical applications [1,2]. For example, protein engineering holds the potential of transforming the metabolic drug landscape through the development of smart, stimulus-responsive drug systems [3]. In industrial processes, protein engineering enables the production of enzymes that offer unique advantages compared with chemical catalysts, such as biodegradability, stereoselectivity, substrate specificity, functionality under relatively mild solvents, temperatures, pHs and pressures, and stability at extreme conditions [4,5]. Today, protein engineering also plays a critical role in advancing the emerging field of synthetic biology, including optimizing pathway enzymes and regulatory elements, balancing pathway redox equivalents, as well as tuning the expression of pathway genes.

There are two main strategies in protein engineering: rational design and directed evolution. Rational design, which is mostly carried out in silico, is knowledge-based, deterministic engineering of proteins. Thus, it needs prior information on the target protein such as a structural model, sequence relationship to homologs, and insights into its biophysical function. As a powerful approach to test general theories in protein chemistry, rational design can be achieved either by single-point mutation, exchange of elements of secondary structure, exchange of whole domains, or by fusion of enzymes. Novel computational tools have constantly improved, resulting in dramatic increase in the sizes of mutant libraries that can be designed. Genome-scale metabolic models (GEMs) have informed and expanded variant design for many industrially relevant microorganisms [6]. Furthermore, multiplex automated genome evolution (MAGE) [7] and CRISPR/Cas [8,9] systems have significantly improved the throughput of genome editing with precision and reduced cost and time required to explore many protein targets. For example, MAGE enables the rapid generation of billions of mutants by repeated insertion, deletion or mutation of DNA at multiple chromosomal targets [7]. Overall, the unprecedented capability of designing and building a large number of variants for rational protein design has placed increased demands on the throughput of screening and selection [10,11].

Directed evolution, on the other hand, mimics the process of natural selection through random mutagenesis to steer proteins or nucleic acids toward desired traits [12–14]. Unlike rational protein design, directed evolution requires neither prior knowledge of a protein's detailed structure nor prediction of the effects of various mutations [15]. Indeed, directed evolution makes it possible to identify undiscovered protein sequences which have novel functions. Moreover, synthetic biologists also increasingly rely on directed evolution to optimize engineered biological systems [16–18]. However, for directed evolution to be truly effective, very large libraries of mutants must be screened under conditions that closely match the desired functionality [19]. Thus, the throughput of screening for variants with improved traits is a major factor dictating the efficiency of directed evolution given that the libraries of random mutants can be easily on the scale of 10^8–10^9 [19–21].

Proteins are engineered for properties such as affinity, selectivity, stability or enzymatic activity [22]. Often, when a cell based fluorescent readout is achievable, screening of engineered protein libraries is performed with fluorescence-activated cell sorting (FACS) to separate a population of cells into sub-populations based on fluorescent labeling [22]. In this case, the phenotype-genotype connection is unbroken because the cells are selected directly. For example, Lipovšek et al. [23] reported in vitro selection of catalytically active enzymes (horseradish peroxidase) from large libraries of variants displayed on the surface of the yeast *S. cerevisiae* and separated by FACS. To improve substrate specificity of glycosyltransferases by directed evolution, Aharoni et al. [24] screened a library of over a million sialyltransferases mutants using FACS and found a variant with up to 400-fold higher catalytic efficiency for transfer to a variety of fluorescently labeled acceptor sugars. In their study, the formation of sialosides in intact *E. coli* cells was detected by selectively trapping the fluorescently labeled transfer products within the cell and the resulting cell population was analyzed and sorted using FACS [24]. However, many desirable properties are not amenable to direct interrogation via FACS because the phenotype is not inherent in a single cell, for example, when improving a protein excreted into growth medium. Properties such as extracellular analyte consumption, product secretion and cell-cell interactions are not readily detectable with flow cytometry. Screening for 'non-cellular' phenotypes necessitates the compartmentalization of single cells or an alternative expression system to maintain the linkage between the phenotype that the selection acts on and the genotype in which the evolutionary information is encoded [19]. Compartmentalization of assays in arrays of wells makes microtiter plates by far the most widely used screening platform. However, the microplate-based method becomes problematic when the assay volume is less than 1 µL due to evaporation and capillary forces [25]. Even with robotic automation for liquid handling using 1536-well plates and assuming a processing rate of 1 plate per minute, the throughput of a well based assay is approximately 25 samples/sec for a prompt optical measurement.

Miniaturization of screening systems can substantially increase sorting efficiency, improve selection and reduce screening costs, enabling exploration of very large libraries (10^8–10^9). These advantageous properties have stimulated emerging micro- and nanotechnologies to move into applications in the life sciences and molecular biology. Early efforts included the in vitro compartmentalization demonstrated by Tawfik and Griffiths [26] in late 1990s. They showed the selection of genes encoding HaeIII methyltransferase from a 10^7-fold excess of genes encoding another enzyme using water-in-oil emulsions. These polydisperse droplets were generated by adding an in vitro transcription/translation reaction mixture into stirred mineral oil containing surfactants. As many assays require an accurate and reliable means of fluid manipulation to enable reproducible results, polydispserse droplets can be problematic. In the sub-microliter or sub-nanoliter volume range, droplet microfluidics, which emerged at the beginning of 2000s [27], presented a new paradigm for screening, offering precise and reduced reagent volumes as well as single-cell resolution analysis [28]. Microfluidic devices, featuring a network of channels with dimensions from tens to hundreds of micrometers, enable the generation and digital manipulation of droplets of uniform sizes (microliter to femtoliter) at very high throughput (up to several kHz). Surfactant systems enable the stabilization of droplets such that they can be incubated off-chip and reintroduced intact into subsequent microfluidic device(s) for sorting and analysis. However, it was not until the most recent decade that the HTS capacity of droplet microfluidics had been demonstrated for protein engineering, especially directed evolution [19,21,22,29–31].

Here, we review the recent high-throughput screening systems developed for protein engineering that are enabled by droplet microfluidics. The review follows the structure of a typical workflow, as illustrated in Figure 1, which includes the following modules: emulsification, incubation, reagent addition, and sorting. This review article can also serve as a tutorial guide for those who need a droplet-based HTS system for their specific applications but may not have prior knowledge about microfluidics. A number of key challenges and opportunities are outlined in the end to motivate the continued innovation of microfluidics with implications for protein engineering.

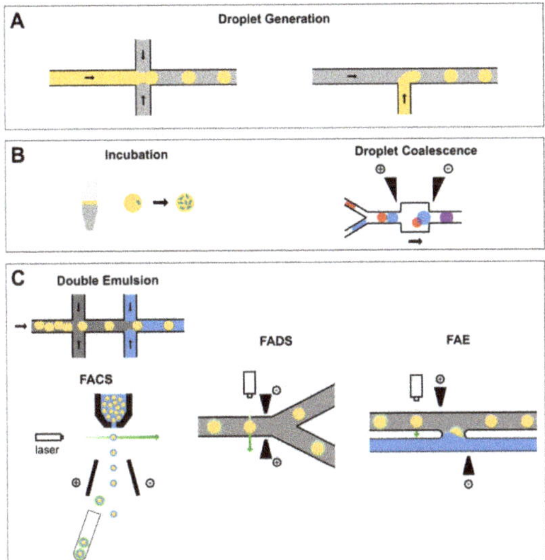

Figure 1. A typical workflow of droplet-based high-throughput screening system for protein engineering. (**A**) Single cells are encapsulated into monodisperse water-in-oil droplets generated on a flow-focusing (left) or T-junction (right) nozzle. (**B**) Cell-laden droplets can be incubated either on-chip or off-chip and new reagents can be added via droplet coalescence. (**C**) Droplets of interest can be sorted either by fluorescence-activated cell sorting (FACS) after an extra step of double emulsification or by fluorescence-activated droplet sorting (FADS). Alternatively, the content of droplets of interest can be recovered via fluorescence-activated electrocoalescence (FAE).

2. Droplet-Based Hight-Throughput Screening

2.1. Microfluidics Droplet Generation

Droplet formation in microfluidic devices utilizes flows of two immiscible liquids that are controlled independently [27]. Most commonly, an aqueous solution is dispersed into a continuous phase of oil. The T-junction and flow-focusing nozzle (Figure 1A) are the most widely adopted geometries of microfluidic channels for generating droplets, where the stream of the dispersed fluid is pinched off by the continuous carrier fluid through shearing to generate monodisperse droplets. The frequency of droplet generation can reach up to 10 kHz and the droplets volumes may range from femtoliters to nanoliters. There have been several reviews available in the literature that provide complete details about droplet generation using microfluidics [22,32–35]. Therefore, the current review only discusses several fundamental elements in droplet generation using microfluidics, in particular, formulations (i.e., oils and surfactants).

As self-contained microreactors, monodisperse water-in-oil droplets have been used predominantly in the miniaturization of biochemical assays. In this scenario, fluorinated oils such as HFE-7500 paired with fluorosurfactants are widely used as the continuous phase [35]. Fluorinated oils provide desirable merits such as low leakage of organic solutes, high solubility of respiratory gases, as well as compatibility with polydimethylsiloxane (PDMS)—one of the most important materials for flow delivery in microfluidic devices. Alternative systems based on hydrocarbon oils, such as mineral oil, have also been successfully used for applications such as PCR [36] and directed evolution of enzymes [37]. In hydrocarbon systems, however, hydrophobic solutes tend to phase partition into the oil which leads to the cross-contamination among droplets, thereby limiting the range of their applications [38]. It is worth mentioning that molecular diffusion is not necessarily excluded in fluorocarbon systems. For example, the exchange of certain fluorophores was also observed even in fluorinated oil (HFE-7500) [39].

The dispersion of an immiscible fluid into another creates a non-equilibrium thermodynamic system [38,40]. The addition of surfactants acts against the homogenization of each fluid in the system by providing an energy barrier for droplet coalescence, yielding a stabilized dispersion in a metastable state [38]. Although fluorinated oils have favorable properties for biochemical assays, only a limited number of surfactants are available for the stabilization of water-oil interfaces in these systems [38]. Perfluoropolyether-polyethylene glycol (PFPE-PEG), a triblock copolymer consisting of fluorinated alkyl domains, is the most commonly used surfactant in fluorinated oils. Its PEG domains are also critical to prevent nonspecific adsorption of proteins, DNA and RNA at the droplet interface. Surfactants compatible with hydrocarbons include sorbitan monooleate (Span 80), silicone-based ABIL EM 90/180, TWEEN 20/80 and so forth.

The oil-surfactant system has to be formulated carefully since it in part determines the capability of microfluidics to generate controlled liquid structures with complex functionalities at the interface for a variety of applications including protein or enzyme engineering. When selecting the combination of oil and surfactant, several factors need to be taken into consideration, including aqueous composition, biocompatibility, molecular exchange between droplets, stability (e.g., tolerance with thermocycling), as well as compatibility with double or multiple emulsification. Table 1 listed the formulations that were frequently used by the droplet-based screening studies discussed in the following sections.

2.2. Reagent Addition in Droplet Flow

Adding reagents into droplets is one of the most important functions in droplet-based microfluidic systems. Additions enable complex biochemical assays, for example, performing multistep reactions that require mixing new reagents at different times [41]. Reagent addition in droplet flow can be achieved by a variety of methods such as droplet coalescence induced by either electric field [42,43], physical constrictions [44], or certain chemicals [45].

Table 1. Oil/surfactant formulations used in the studies that are discussed in this review.

Oil/Surfactant	Applications
HFE-7500/PFPE-PEG	Double emulsion for flow cytometric sorting [21], bacterial microcolonies in gel beads [46], FADS [30,31,47,48], electrocoalescence [49], triple emulsification for reagent addition [50], chemically-induced droplet coalescence (PFB added to induce coalescence) [45], picoinjector [41], multiplexed FADS [51], droplet size-based separation [52]
FC-40/PFPE-PEG	Double emulsion for flow cytometric sorting [21], FADS [19]
HFE-7500/QX200 (Bio-Rad proprietary oil)	2-way or 4-way FADS [53]
HFE-7500 or FC-40/fluorinated silica nanoparticles	Pickering emulsification to mitigate molecular diffusion [54,55]
Perfluorodecalin and perfluorooctanol (7:3 (v/v))	Fluorescence lifetime-activated droplet sorting [56]
Mineral oil/Span 80	Encapsulation of bacteria in agarose microparticles [57], emulsification of Amplex Ultra Red and horseradish peroxidase (specific sugars added to mitigate molecular diffusion) [58]
Mineral oil/Abil EM 90	Emulsification of 3-O-methylfluorescein (BSA added to mitigate molecular diffusion) [59]

Niu et al. [44] developed a physical method to merge aqueous droplets within a microfluidic chamber featuring several rows of pillars separated by distances smaller than the representative droplet dimension. In a typical case (Figure 2A), a droplet entered the chamber, slowed down and stopped. This droplet would then merge with another droplet behind it when the surface tension was overwhelmed by the hydraulic pressure. It should be noted that in the above study hexadecane was used as the continuous phase without the addition of surfactants. Therefore, the efficiency of this physical coalescence method is subjected to validation for other oil/surfactant formulations.

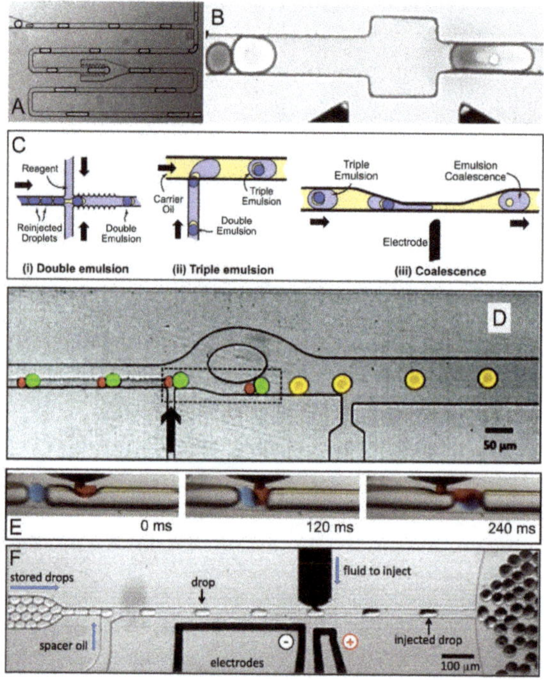

Figure 2. Droplet coalescence approaches. (**A**) The coalescence of two droplets within a pillar constriction-assisted droplet-merging device. Figure 2A was adapted from reference [44] with permission from the Royal Society of Chemistry. (**B**) A fusion module that delivered an alternating current (AC) field permitted electrically controlled merging of pairs of dye-containing droplets and cell-containing droplets. Figure 2B was adapted from reference [60], complying with the License for PNAS Articles. (**C**) Schematic of a different electrocoalescence workflow assisted with triple emulsification. At step (i), reinjected droplets are enveloped by an aqueous reagent phase in a hydrophilic channel. The resulting double emulsion travels to a hydrophobic junction at step (ii) where carrier oil encapsulates it to form a triple emulsion. At step (iii), the encapsulated double emulsion is ruptured in the presence of an electric field. Figure 2C was adapted from reference [50] with permission, copyright American Institute of Physics. (**D**) Passive microfluidic droplet coalescence through the addition of a destabilizing alcohol. Perfluorobutanol is added through the channel indicated by the black arrow, causing downstream coalescence of paired droplets. Droplets are false colored. Figure 2D was adapted from reference [45] with permission from the Royal Society of Chemistry. (**E**) A sequence of still images showing the direct injection of red-dyed octadecene into blue-dyed octadecene droplets, using Ar gas as a spacer and PFPE as the carrier fluid. Figure 2E was adapted from reference [61] under the Creative Commons Attribution license. (**F**) A picoinjector microfluidic device. The spacer adds oil from a side channel to space the droplets. The picoinjector injects fluid by merging the droplets with a pressurized channel containing the reagent. Picoinjection is triggered by an electric field, which is applied by the electrodes. Figure 2F was adapted from reference [41], complying with the License for PNAS Articles.

Electrocoalescence is commonly used to merge pairs of droplets that are synchronized by size-dependent flow in microfluidic channels [42,49,60]. Two streams of different-sized droplets made independently (allowing different time scales, sizes, and compositions) can be synchronized in a single microfluidic channel. Smaller droplets move faster than larger droplets due to the Poiseuille flow. Therefore, the smaller droplets are brought into contact with preceding larger droplets (Figure 2B). Pairs of surfactant-stabilized droplets are coalesced while passing by an electric field. It was found that an electric field that was aligned with the flow direction could induce the maximum force to deform the adjacent surfaces of the paired droplets, thereby maximizing the effectiveness of electrocoalescence [62]. Alternatively, Sciambi and Abate [50] presented another method to add reagent to droplets by enveloping them as double emulsions in reagent-filled droplets and then rupturing the inner droplet with an electric field (Figure 2C). When the double emulsions ruptured, they released their contents into the enveloping droplets, ensuring mixing with reagent while limiting cross-droplet contamination.

In addition, Akartuna et al. [45] presented a chemically-mediated approach to coalesce pairs of surfactant-stabilized water-in-fluorocarbon oil droplets in a continuous flow (Figure 2D). Pairs of large and small droplets were exposed to perfluorobutanol, a poor solvent for the surfactant, before entering into a constriction channel. In the presence of perfluorobutanol, the water-oil interfaces became highly prone to coalescence due to the local depletion of surfactant. When the pairs exited the constriction, the velocity of the leading large droplet rapidly decreased, forcing the droplet pairs to contact and coalesce before re-stabilization. Overall, this coalescence method was able to perform ~300 droplet-merging events per second on a par with the rates of electrocoalescence methods [19,42,60], although it may be majorly limited by upstream pairing.

The coalescence-based methods discussed above involve an additional stream of droplets which encapsulate the reagent(s) to be added. The reagent droplets have to be synchronized with the target droplets such that they can be brought to close contact in pairs and merged via constriction-, chemical- or electric field-induced coalescence. The major advantage of coalescence-based methods is a broad range of reagent volume that can be added, especially for large-volume addition. Electrocoalescence also works compatibly with stable emulsions. On the downside, these methods, which rely on droplet pairing, tend to be inconvenient to conduct multiple additions since the synchronization of several streams of droplets becomes necessary but technically challenging [41]. Additionally, the electric field or chemicals used to induce coalescence may not be biocompatible.

In contrast to the paired droplet methods, a T-junction can be used to introduce a reagent as a continuous stream. The reagent is injected directly from the side channel into the droplets that flow in the main stream when they pass the T-junction [61,63,64]. In this approach, the reagent stream needs to flow much slower than the main stream to avoid the formation of new reagent only droplets within the main stream. To mitigate excess reagent flow, several methods are available to increase the efficiency of the direct injection method, such as the addition of inert gas spacers or a double T-junction structure. For instance, Nightingale et al. [61] used argon gas plugs to maintain uniform droplet spacing and mitigate the undesired formation of reagent only droplets. They employed the improved direct injection approach to conduct a five-stage quantum dot synthesis wherein particle growth was sustained by repeatedly adding fresh feedstock [61] as seen in Figure 2E.

One of the limitations of direct injection is the incompetency of injecting reagent into stable emulsions since the surfactants that are used to stabilize the target droplets prevent reagents from entering them. By leveraging electrocoalescence, Abate et al. [41] developed a picoinjector which used a pressurized channel to inject a controlled volume of reagent into preformed, PFPE-PEG-stabilized droplets at kHz frequency, as shown in Figure 2F. The injection volume can be controlled with sub-picoliter precision by tuning the droplet velocity and injection pressure. More importantly, they established serial and combinatorial injections by laying out several picoinjectors sequentially, each separately controlled by an electric field [41].

2.3. Flow Cytometric Sorting of Droplets

Commercially available fluorescent activated cell sorters are designed to interrogate suspended cells at ~10^7 per hour and select the "hits" from a heterogeneous mixture based on the specific fluorescent and light scattering properties of the sample. FACS is, however, incompatible with non-aqueous carrier fluids. Thus, double emulsification of water-in-oil-in-water and other techniques become a plausible approach to enable the flow cytometric sorting of droplets, providing FACS compatible compartments to maintain the genotype-phenotype linkage in an aqueous carrier stream.

In vitro compartmentalization in double emulsions for FACS was demonstrated based on fluorescent signals in early 2000s [65–67]. However, double emulsions used in these pioneer studies were generated by homogenization with high polydispersity, which potentially limited the sensitivity and the throughput of FACS [29]. Recently, Terekhov et al. [29] combined microfluidic production of double emulsions with efficient FACS selection, as shown in Figure 3A,B. With integration with next-generation sequencing and liquid chromatography-mass spectrometry, their study provided deep insights into the genotype-phenotype linkages of the secretomes of encapsulated organisms. The functionality and versatility of the platform were demonstrated with the selection of different biocatalytic activities, screening enzymes with different levels of the same activity, de novo creation of enzymes with artificial activity, and investigation of bacterial cell-to-cell interactions [29].

Zinchenko et al. [21] also established a protocol with complex elements for quantitative analysis and sorting of monodisperse double emulsion droplets in a commercial flow cytometer. Their workflow (Figure 3C) incorporated additional steps, including assaying heat inactivation of lysates within the droplets, concentration of the encapsulated contents induced by droplet osmosis, and storage of droplets at −80 °C for discontinuous workflows. In their study, single *E. coli* cells expressing either wild-type arylsulfatase from *Pseudomonas aeruginosa* or a low activity variant (H211A) were initially co-compartmentalized with lysis reagents and enzyme substrate in water-in-oil droplets forming a primary emulsion (Figure 3Ci–iii). The primary emulsion droplets were then introduced into a second microfluidic device for double emulsification (Figure 3Civ–vii). The employment of two separate devices to produce double emulsion droplets rendered the independent size control of primary and double emulsions and the flexibility of surface coating (i.e., fluorophilic coating for primary emulsion and hydrophilic coating for double emulsion) [21]. The double emulsion droplets of interest contained active enzyme that was expressed by transformed *E. coli* and released due to cell lysis. The active enzyme hydrolyzed the substrate fluorescein disulfate and generated a fluorescent readout of enzyme activity. The droplets containing H211A variant only showed a low level of background fluorescence (due to the lack of hydrolytic activity). Finally, the highly fluorescent population was sorted via FACS to obtain the active variants. The results demonstrated that a sample with 0.01% active cells in the initial population was enriched 2500-fold. Further, 100,000-fold enrichment was achieved for another sample that initially contained only one hit in 1,000,000 cells (0.0001% cells expressing active enzyme), although with a higher droplet occupancy [21].

Eun et al. [57] described an HTS method for isolating spontaneous mutants of *E. coli* that had developed resistance to the antibiotic rifampicin using FACS. Instead of being compartmentalized into double emulsions, *E. coli* cells (MG1655-p*tet*EGFP) were encapsulated in agarose microparticles on a flow-focusing microfluidic device and incubated in the presence of varying concentrations of rifampicin. Microparticles containing Green Fluorescent Protein (GFP)-positive cells (0.2% of the total population) were sorted on FACS. The mutants were recovered from the microparticles and sequenced. As a result, an A1538T base-pair mutation was identified in the *rpoB* open reading frame. This mutation resulted in a Q513L substitution, conferring resistance to rifampicin. The overall screening consumed a total of 65 μg of rifampicin and took 6 h to complete, compared to 15 mg of compound that would have been required for a 48-h screening based on agar plates [57].

By performing FACS to sort cell-laden hydrogel microparticles, Duarte et al. [46] presented a high-throughput screening method to address challenges like substantial cell-to-cell variability and the requirement to check multiple states in synthetic biology (Figure 3D). In their study, single *E. coli* cells, either expressing or not expressing superfolder GFP, were encapsulated in water-in-oil droplets (20 μm in diameter), in which the aqueous phase, containing 1% agarose, solidified to a gel when chilled. FACS selection was performed after sufficient time to allow formation of microcolonies and expression of GFP. Sorted cells were recovered by enzymatic digestion of the agarose and plated on agar plates to determine the enrichment rate. The results showed that an enrichment of 30,000-fold was achieved for a 1:100,000 dilution (0.001%) of GFP expressing versus non-GFP expressing cells, which is comparable to enrichment rates previously reported [21,29,68].

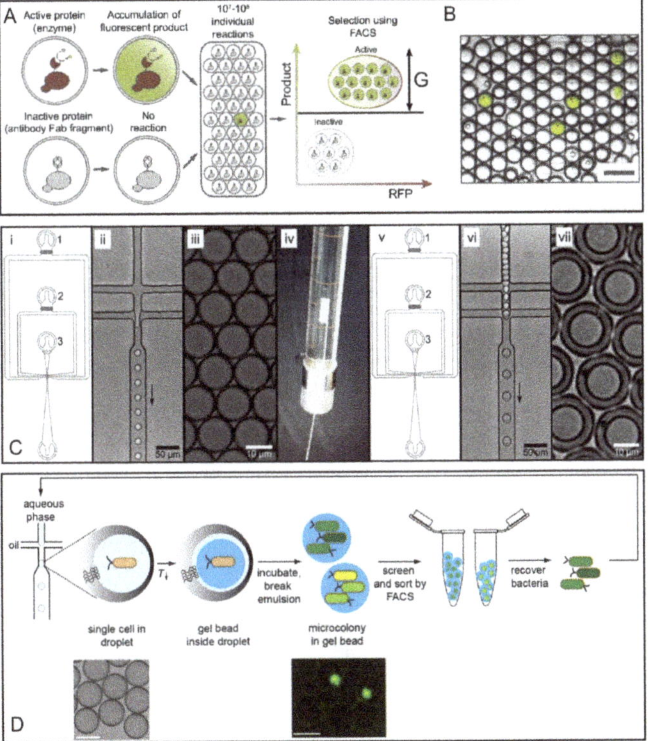

Figure 3. Flow cytometric sorting workflow using either double emulsion or hydrogel encapsulation. (**A**,**B**) Screening of biocatalysts anchored to the yeast surface using FACS on water-in-oil-in-water emulsions. (**A**) Compartmentalization of active and inactive yeast cells with fluorogenic substrate. After the mixture of active and inactive cells was encapsulated, the fluorescent product accumulated solely inside the droplets with active cells, which were selected using FACS. (**B**) Visualization of biochemical reaction by merging the signals of green fluorescence (reaction product), red fluorescence (reporter protein), and the visible light image. (Scale bar, 100 μm.) Figure 3A,B were adapted from Reference [29], complying with the License for PNAS Articles. (**C**) Formation of double emulsion droplets using a two-chip system. (i) Design of the device for primary emulsification; (ii) The aqueous samples are first mixed, then primary droplets are formed in the flow-focusing junction; (iii) Image of the monodisperse water-in-oil droplets; (iv) The emulsion droplets are taken up in a syringe, overlaid with mineral oil, and cushioned with a bottom layer of fluorinated oil; (v) A device with identical design to the first emulsification device, but a hydrophilic coating is used for formation of secondary emulsions; (vi) Image showing the production of water-in-oil-in-water double emulsion; (vii) Image of the monodisperse double emulsion droplets. Figure 3C was adapted from reference [21] under the Creative Commons Attribution license. (**D**) Overview of the FACS screening method with hydrogel encapsulation. Single cells are encapsulated into monodisperse water-in-oil emulsion droplets. The aqueous solution contains agarose that gels upon cooling on ice. After the formation of monoclonal microcolonies inside the beads, the beads are recovered from the emulsion and sorted by FACS. The bacteria are recovered from the gel beads and are then ready for a further round of analysis. Bottom left: a phase contrast microscope image of droplets (bottom left). Bottom right: a fluorescence microscope image of beads with two of them containing a microcolony. Scale bars: 50 μm. Figure 3D was adapted from reference [46] with permission from the American Chemical Society.

Hydrogels are cross-linked hydrophilic polymers swollen in an aqueous environment [69]. Solute transport within hydrogels occurs primarily in the water-filled regions delineated by the polymer chains [69]. Therefore, cell-laden hydrogel microparticles may not be sufficient to prevent small molecules like substrates from diffusing between microparticles. However, the tunability of the physical and chemical parameters of hydrogels offers the opportunity of regulating the diffusion of solutes between microparticles in a selective manner. In other words, the structure and density of the crosslinking agent used in the formulation will affect the network mesh size, polymer chain mobility and the charge of polymer chains, and thus the rate of the diffusion of water-soluble molecules within and out of the hydrogel matrix [70]. Additionally, due to their mechanical properties, hydrogel particles can tolerate more shear and experience less deformation while they travel through microscale channels, thereby mitigating accidental rupture and coalescence. It has also been demonstrated with mammalian cells that oxidative stress was reduced by coating the cell-seeded gelatin core with a biodegradable silica shell without compromising the transport of nutrients to the cells encapsulated [71].

2.4. Fluorescence-Activated Droplet Sorting

Although in vitro compartmentalization can increase the versatility of FACS, flow cytometric sorting of double emulsions or hydrogel encapsulation still suffers from several limitations. For instance, the generation of double emulsions or hydrogel encapsulation needs additional steps of emulsification or microgel gelation/recovery, presenting additional complexity to the already lengthy droplet sorting process. It is also less convenient to manipulate the content of microcompartments like double emulsions after encapsulation.

To address this challenge, Baret et al. [30] developed a fluorescence-activated droplet sorting (FADS) system that was optimized to directly sort picoliter-sized droplets by dielectrophoresis. Water-in-oil droplets were introduced into a microfluidic sorting device where they were spaced out and sorted at an asymmetric Y-shaped junction (Figure 4A–C). Droplets flowed along the wider 'negative' arm of the sorting junction by default due to the lower hydraulic resistance. If a droplet were to be sorted, based on a fluorescent signal for example, a pulse of high voltage alternating current would be applied across the electrodes adjacent to the sorting junction. The resulting electric field deflected the droplet of interest into the narrower 'positive' arm of the junction by dielectrophoresis [72]. To validate the system, *E. coli* cells, expressing either the reporter enzyme β-galactosidase or an inactive variant, were compartmentalized with fluorogenic substrates within 12-pL monodisperse droplets and sorted at a rate of ~300 droplets per second. The false positive (i.e., droplets that were sorted but contained inactive variant) error rate of the sorting device at this throughput was less than 1 in 10,000 droplets. When the cells were encapsulated at a lower density (e.g., ~1 cell for every 50 droplets), all of the sorted and recovered cells were the active strain. An even higher throughput (~2000 droplets per second) may be feasible but at the cost of an increase in the false positive error rate (e.g., 1 in 100 droplets).

Agresti et al. [19] explored the application of FADS in aiding the discovery of variants of the enzyme horseradish peroxidase (HRP) in directed evolution as shown in Figure 4D,E. Yeast cells (*Saccharomyces cerevisiae* EBY100) displaying enzyme variants on their surfaces were co-encapsulated into 6-pL aqueous droplets with a fluorogenic substrate. To ensure single-cell encapsulation and avoid confounding multiple cells, only ~22% droplets contained cells according to the Poisson distribution. As the droplets flowed through a FADS device, those containing the most active enzyme variants were sorted based on the fluorescence intensity. In each round of screening, the yeast cells were induced to express HRP, encapsulated into droplets and sorted at 2000 droplets per second for up to 3 h, with a total of up to 2×10^7 cells being interrogated. As a result, new mutants of the enzyme HRP were identified exhibiting catalytic rates more than 10 times faster than their parent. In total, ~10^8 individual enzyme reactions were screened within 10 h, consuming less than 150 µL reagent. Compared with the state-of-the-art robotic screening systems, it was estimated that the FADS-based high-throughput screening system performed the entire assay with a 1000-fold increase in throughput and a 1-million-fold reduction assay reagent cost.

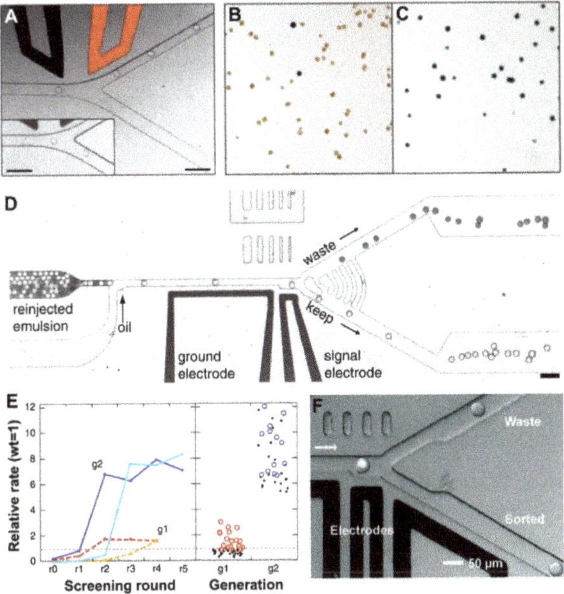

Figure 4. Fluorescence-activated droplet sorting systems. (**A**) Trajectories of droplets flowing through the sorting junction of a FADS system. When an AC electric field is applied across the electrodes, the droplets are deflected into the 'positive' arm. In the absence of the electric field, the droplets flowed into the 'negative' arm by default (inset). (**B,C**) Enrichment of cells by FADS based on β-galactosidase activity. Photographs of *E. coli* colonies before (**B**) and after sorting (**C**). The *lacZ* bacteria (blue colonies) were completely purified from the Δ*lacZ* bacteria (white colonies) after sorting, resulting in only *lacZ* colonies growing on the agar. Figure 3A–C were adapted from Reference [30] with permission from the Royal Society of Chemistry. (**D**) A different design of the FADS system. The droplets flow as a solid plug to a junction where oil is added to space the droplets. Light droplets contain 1 mM fluorescein and the dark ones contain 1% bromophenol blue. When a droplet passes by the laser that is focused on the channel at the gap between two electrodes, its fluorescence intensity is detected. If the intensity is above a threshold (in the case of light droplets), the droplet is sorted by dielectrophoresis towards the bottom channel. (**E**) Enrichment of library pools. The activities are normalized relative to wild-type HRP. The first-generation epPCR and saturation mutagenesis libraries (dashed red and orange, respectively) enrich to a level of about two times the activity of the wild type after four sorting rounds. The second-generation low- and high-mutation rate libraries (solid blue and cyan, respectively) enrich to about eight times the wild type. The right panel shows a dot plot of the activities of the 50 unique first-generation (g1) mutants (red circles) and 31 second-generation (g2) mutants (blue circles). Figure 3D,E were adapted from reference [19], complying with the License for PNAS Articles. (**F**) The sorting junction of another FADS system. Single droplet fluorescence was detected following excitation by the laser (the white dot). The default flow path of the droplet is towards the top 'waste' channel since the waste outlet is at atmospheric pressure and a withdrawal of less than half the total flow rate is applied to the 'sorted' outlet. However, if the droplet fluorescence exceeds a predefined threshold an electric field is activated between the electrodes, pulling the droplet to the bottom channel. Figure 3F was adapted from Reference [47] with permission from the Royal Society of Chemistry.

Sjostrom et al. [47] used a FADS system to screen a yeast library with mutations randomly introduced throughout the genome by UV-irradiation mutagenesis to select the cells with high α-amylase production (Figure 4F). A total of ~3 × 10⁶ droplets were sorted at a rate of 323 droplets per second over the course of slightly more than 2 h. They found that cells from the sorted subpopulation had over 60% higher enzyme production and 35% higher yield than the unsorted library. The top-performing clone was found to double the enzyme production compared with the mother strain. As suggested in this study, the platform had a throughput over 300 times higher than an automated microtiter plate screening system. At the same time, reagent consumption for a screening experiment was decreased by a million-fold, greatly reducing the costs of evolutionary engineering of production strains.

Similarly, Ma et al. [48] described a high-throughput microdroplet sorting system that could be used to screen up to ~10^7 enzyme variants (10^8 droplets) per day. The proposed system was able to evaluate two reaction channels simultaneously by adopting a dual-fluorescence detection/sorting microfluidic device. The employment of different combinations of two-color fluorogenic substrates enabled the screening for enzyme variants that had both improved catalytic activity and an additional enzymatic property such as regioselectivity, chemoselectivity, or enantioselectivity. As an example, Ma et al. [48] used the system to engineer the enantioselectivity of an esterase to preferentially produce desired enantiomers of profens, an important class of anti-inflammatory drugs. Using two types of selection modes over the course of five rounds of directed evolution, they identified a variant with 700-fold improved enantioselectivity for the desired (*S*)-profens from 5 million mutants.

2.5. Fluorescence-Activated Electrocoalescence

Instead of sorting intact droplets containing "hits", Fidalgo and coworkers [73] presented a different approach to selectively incorporate the contents of droplets initially flowing in a carrier oil phase into a continuous aqueous stream by electrocoalescence (Figure 5A). As soon as the fluorescence intensity of a droplet met a predefined criterion, an electric field across the channel was triggered to force the droplet of interest to cross the oil-water interface and coalesce with an aqueous stream at a frequency of 10–250 Hz. In their study, a mixture of fluorinated oil (FC-77) containing 30% (v/v) perfluorooctanol was employed as the carrier phase to compromise droplet stability and facilitate coalescence. Fluorescence-activated electrocoalescence has great potential of integrating with existing microfluidic modules for droplet-based screening. Additionally, the composition of the receptor stream can be adjusted to suit specific applications, such as quenching reactions to reach well-determined endpoints, providing favorable culture conditions for extracted cells, and lysing cells for downstream analysis [73]. However, the sorting method is destructive to the droplets, the content of the selected droplets is released into the aqueous stream via coalescence and the sorted droplets no longer exist.

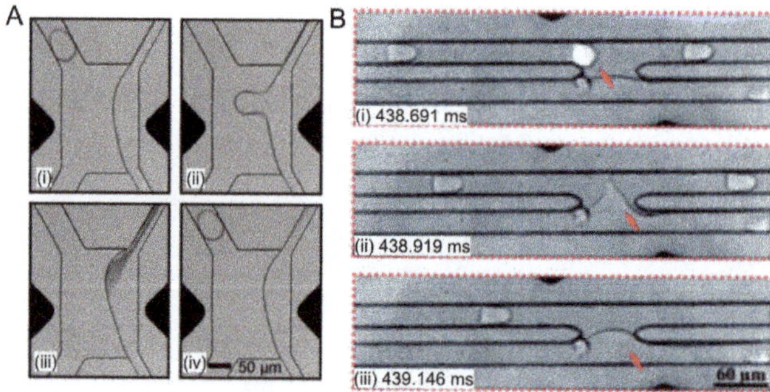

Figure 5. Fluorescence-activated electrocoalescence. (**A**) Selective extraction of the content of droplets of interest. (i) Below the threshold voltage, the electric field is insufficient to induce coalescence. As a result, the droplet passes by without coalescence. (ii and iii) When an additional square pulse is applied, an individual droplet is selected, and its contents are incorporated to the lateral aqueous stream. (iv) The applied voltage returns to its original value before the next droplet enters the electrode region and therefore the droplet flows past without coalescence. Figure 4A was adapted from Reference [73] with permission from John Wiley & Sons, Inc. (**B**) The fluorescence-activated electrocoalescence of a single droplet (indicated by the red arrow) in another FAE device. Figure 4B was adapted from Reference [49] with permission from the Royal Society of Chemistry.

The above electrocoalescence-facilitated droplet 'sorting' method was further optimized by Fallah-Araghi et al. [49] to achieve reliable sorting of up to 2000 droplets per second. In the design shown in Figure 5B, the gap in the coalescence section was minimized to allow efficient droplet extraction while maintaining a stable interface between the aqueous stream and the oil stream in the absence of an electrical field. They assembled the modules of encapsulation, droplet fusion, size-based droplet sorting and fluorescence-activated electrocoalescence to perform in vitro ultrahigh-throughput screening for protein engineering and directed evolution. Single genes were compartmentalized in aqueous droplets. After amplification by PCR, the droplets containing 30,000 copies of each gene were paired and fused with droplets containing a cell-free coupled transcription-translation (IVTT) system and the reagents for a fluorogenic assay. Then, genes from droplets containing desired activities were selectively recovered using fluorescence-activated electrocoalescence. As a result, a 502-fold enrichment was achieved from a mixture of *lacZ* genes encoding β-galactosidase and *lacZmut* genes encoding an inactive variant (molar ratio 1:100). Meanwhile, the volume and cost of PCR and IVTT reagents were reduced by ~105-fold compared with microtiter plate-based methods, achieving the screening of 10^6 genes at the cost of 150 μL of reagents. In summary, we have compared the key advantages and disadvantages of the three main droplet sorting strategies in Table 2.

Table 2. Advantages and disadvantages of existing droplet-based HTS technologies.

Droplet Sorting Mechanism	Advantages	Disadvantages
Flow cytometric sorting	Compatible with commercial instruments	Needs additional emulsification or hydrogel gelation; Droplet size is restricted by instrument specification
Fluorescence-activated droplet sorting (FADS)	Single emulsification; Sorted droplets are available for more rounds of sorting; Wide range of droplet sizes	Requires careful balance of "collection" and "waste" fluid flows; Needs additional step to recover the content inside the droplets
Fluorescence-activated electrocoalescence	Direct recovery of droplet content; Tunable conditions of the collection aqueous stream to perform additional chemistry	Sorted droplets no longer exist; Not possible to perform second round of sorting

2.6. Mitigating Molecular Diffusion between Droplets

For droplets to be self-contained, a variety of surfactants have been explored to stabilize the water-oil interface and prevent droplet coalescence. Unfortunately, it has been found that many of the commonly used surfactants can actually assist molecular diffusion between droplets, resulting in cross-contamination of small, hydrophobic fluorescent molecules [39,58,59,74,75]. The leakage of fluorescent probes across droplet interface can be attributed to several phenomena. Fluorescent molecules may directly diffuse into the continuous phase [55,59]. Alternatively, micelles formed by surfactant molecules may facilitate the leakage of fluorescent probes [39,59]. In the latter case, a surfactant concentration below the critical micelle concentration would help mitigate the leakage but the droplets would become prone to coalescence if a low concentration of surfactants is used [55].

Since high-sensitivity assays are typically based on fluorescent readout, fluorophore exchange among aqueous droplets must be avoided to reliably compartmentalize the genotype-phenotype linkage. Courtois et al. [59] found that the addition of 5% (w/v) bovine serum albumin (BSA) significantly reduced the leakage of 3-*O*-methylfluorescein to 3%, compared with 45% for mineral oil containing 0.75% (w/w) Abil EM 90 surfactant. Moreover, Sandoz et al. [58] reported that the addition of 25% sucrose mitigated the leakage of the dye resorufin when mineral oil was mixed with different concentrations of surfactants combining 9% (w/w) Span 80 and 1% (w/w) TWEEN 80. They suggested that the sucrose strongly interfered with the surfactant's ability to form micelles in water.

In fact, emulsion stability does not necessarily require amphiphilic surfactants to reduce the interfacial tension. The original observations by Ramsden [76] and the seminal work by Pickering [77] demonstrated that emulsion stability could be efficiently promoted by dispersed particles in the colloidal size range, which is now known as a Pickering emulsion. Prior work on making nanoparticle-stabilized emulsions in microfluidic systems mostly used hydrocarbons as the continuous phase [78–80]. Pan et al. [55], however, designed and synthesized silica nanoparticles (F-SiO$_2$ NPs) and generated a Pickering emulsion in fluorinated oil. These nanoparticles are irreversibly adsorbed to the water-oil interface and do not form surfactant micelles which are likely responsible for the leakage of small molecules among aqueous droplets. In their study, the leakage of resorufin was prevented when the droplets were stabilized by F-SiO$_2$ NPs. However, enzymes and proteins that adsorb nonspecifically to silica surfaces can become denatured. Thus, fluorinated silica nanoparticles could reduce enzyme activities and compromise the performance of droplet-based enzymatic assays [81,82]. To address this issue, the above Pickering emulsion system was improved by introducing PEG into the dispersed phase such that PEG could adsorb onto the surface of the F-SiO$_2$ NPs from within the droplets [54]. The adsorption of PEG is driven by the formation of hydrogen bonds between PEG and the silanol groups on the nanoparticle surfaces [83]. PEG-adsorbed F-SiO$_2$ NPs were found to be effective both in preserving enzyme activity and in preventing the leakage of small molecules. Although NPs are usually attached to the droplet surface more strongly than traditional surfactants due to the high desorption energy of NPs from the droplet surface, Pan et al. [84] showed that methods like electrocoalescence and perfluorooctanol addition worked well for merging NP-stabilized droplets with high efficiency to recover the contents of Pickering emulsions.

Instead of inhibiting molecular diffusion chemically, Siltanen et al. [85] constructed defined reactions with chemicals and cells incubated under air on an open array, thereby physically isolating the droplets and retaining hydrophobic compounds in their compartmentalized reactors. This oil-free picodrop bioassay platform was built upon the so-called Printed Droplet Microfluidics (PDM) technology that was developed by the same group in 2017 [86]. The ordered array of droplets generated by PDM functions analogously to microliter well plates—but at one-thousandth the scale—and uses a "microfluidic robot" for deterministic and programmable reagent and cell dispensing [86]. The PDM platform was used to detect the small differences in *trans*-β-farnesene production between similar strains of yeast. Farnesene is a sesquiterpene of the mevalonate and deoxyxylulose-5-phosphate pathways and readily soluble in emulsion oil. By compartmentalizing reactions under air, farnesene molecules remained partitioned with the cells responsible for their production, allowing direct measurements of product concentration for screening. The PDM platforms can interface with most bioanalytical tools and combines the throughput and low reagent consumption of droplet microfluidics with the flexibility and control of robotic fluid handling of physical arrays [85].

3. Commercialization and Commercial Activities of Droplet-based HTS in Protein Engineering

The protein engineering market is rapidly growing and promises lucrative opportunities given the increasing number of proteins and enzymes engineered for a multitude of applications in food, biopharmaceutical, and environmental industries. In fact, the commercialization of engineered proteins has manifested in economically beneficial and viable solutions for industry and healthcare sector. Protein engineering has also evolved to become a powerful tool contributing significantly to the developments in synthetic biology and metabolic engineering [87]. According to the report published by Zion Market Research in early 2019 [88], the continued expansion of the global protein engineering market is likely to be majorly driven by continuously developing pharmaceutical industries, biotechnological innovations, increasing focus toward targeting specific drug development, rising clinical and analytical techniques, and a substantial increase in research funding over the estimated timeframe. The report suggests that the global protein engineering market was valued at approximately USD 1.1 billion in 2018 and is projected to generate around USD 2.6 billion by 2024, at a compound annual growth rate (CAGR) of about 15.5% between 2019 and 2024. However, the costly and technologically complex tools and systems required for protein engineering, dearth of skilled labor, and need for highly qualified researchers pose as a hinderance to the growth of this market.

These challenges can be addressed by leveraging emerging technologies such as advanced computational protein design using artificial intelligence and machine learning and novel high-throughput screening enabled by miniature technologies like microfluidics. Droplet-based HTS technology has the potential to save considerable assay time and reagent cost, especially when exploring large libraries such as those from directed evolution. Droplet generation and cell encapsulation can now be performed readily on commercial microchips provided by microfluidics companies like Fluigent and Dolomite. Further, there are an increasing number of biotech companies applying or developing droplet-based microfluidics to accelerate the screening and selection for protein engineering. In a pilot study, researchers led by Merck coupled droplet microfluidics with mass spectrometry to accelerate the screening of enzyme evolution libraries for transaminase [89]. Velabs Therapeutics, a spin-off company launched by the European Molecular Biology Laboratory (EMBL), has been pioneering the screening of antibodies with modulatory function on complex signaling proteins using a fully integrated droplet-based microfluidics screening platform for phenotypic assays. Their system incorporates reinjection, fusion and sorting of droplets on a single microchip [90]. Additionally, Biomillenia and Merck are collaborating on the development of a market-leading enzyme in Merck's portfolio of industry-leading enzyme solutions by utilizing Biomillenia's droplet microfluidics screening technology. In a collaboration with Soufflet Biotechnologies, Biomillenia's droplet screening platform was also used to develop novel cellulolytic and proteolytic enzymes as additives to animal feed, resulting in enhanced growth and expanding the range of raw materials for feed formulations. In addition, Hooke Bio, an Irish biotech company, developed a droplet-based, cell-screening device called Enigma which is highly adaptable for a multitude of applications including combinatorial drug screening. The Enigma device utilizes water-in-oil-in-water emulsions to allow diffusion of nutrients across the oil membrane and removal of waste products, which enables long-term (>48 h) cell culture for certain disease models such as type 1 diabetes. Sphere Fluidics provides an integrated platform (Cyto-Mine) to streamline cell encapsulation, assay, and FADS-based droplet sorting to accelerate the screening of antibiotic-resistant bacteria [91], antibody discovery and cell line development.

4. Challenges and Opportunities Facing Droplet Microfluidics in Protein Engineering

The past decade has witnessed a growing number of droplet-based HTS systems that accelerated the "design-build-test" cycles in the rational design and/or directed evolution of proteins. These exciting advancements also provided compelling evidence that droplet microfluidics can facilitate automation and enable new assay modalities for high-throughput screening. Nevertheless, there are still several potential challenges that need to be addressed. These challenges also represent the opportunities for continued innovation of microfluidics with implications for protein engineering.

(1) Most FADS systems rely on a lower hydraulic resistance along the 'negative' arm of the sorting junction to prevent droplets from entering the 'positive' arm in the absence of dielectrophoretic forces. Such a delicate hydraulic balance is very sensitive to the fluctuation of flow rates at the sorting junction. Fluctuations of fluid dynamics can be caused by the pulsing of syringe pumps, droplet aggregates, and/or the accumulation of precipitates during lengthy screening applications [53]. Therefore, more robust mechanisms to regulate the default flow path of droplets would be highly useful, especially for scaled-up, industrial applications. For example, multiple electrode pairs can be arranged parallel to the channels to achieve highly reliable two-way sorting, largely independent of the relative flow rates in the channels downstream of the sorting junction [53].

(2) Current droplet sorting systems based on water-in-oil emulsions almost always perform binary sorting. High-throughput multiplexed sorting capabilities would permit droplet sorting systems to sort based on multiplex signals more in line with flow cytometry. By integrating binary sorters serially, the sorting of three different droplet populations can be achieved, although the throughput was limited to 2–3 droplets per second [53]. Another FADS system was developed to sort up to five

different droplet populations simultaneously at rates of several hundreds of droplets per second [51]. Multiplexed FADS systems like these will have the potential of significantly expanding the scope of how droplet-based HTS can accelerate the iterations for protein engineering.

(3) Variability in the number of cells per droplet due to stochastic cell loading is a major barrier to the encapsulation of cells within picoliter-sized monodisperse droplets. Dictated by the Poisson statistics, very low average loading densities have to be used to ensure single-cell encapsulation and avoid confounding multiple cells, which means that most droplets actually contain no cells and these 'meaningless' droplets consume the majority of the screening capacity. Approaches that can increase the single-cell occupancy can significantly increase the efficiency of droplet-based screening. Previously, inertial focusing was employed to evenly space cells as they travel rapidly within a high aspect ratio microchannel such that cells could enter the flow-focusing nozzle with the same frequency as droplet formation [92,93]. In the proof-of-concept experiment, over 80% droplets contained a single particle (i.e., 9.9 μm-diameter polystyrene beads), compared with about 40% from the Poisson distribution. Unfortunately, the technology of inertial focusing has been largely limited to bioparticles larger than red blood cells due to the strong correlation between the inertial lift forces and the particle size [94]. Very recently, Cruz et al. [95] has demonstrated inertial focusing in curved channels and the alignment of particles with diameters of 0.5–2 μm, a range that comprises a multitude of bacteria and yeast cells, which could be integrated with existing droplet-based HTS assays for protein engineering.

(4) Current microfluidic droplet sorting systems trigger the dielectrophoretic force based on fluorescence intensity, which renders these techniques high sensitivity and high throughput. However, fluorescence intensity can be biased by many factors such as background fluorescence, the dye concentration, the intensity of the light source, quenching and light scattering [56]. Thus, it would be highly desirable to explore droplet sorting systems, either active or passive, that are based on other properties such as average fluorescence lifetimes [56], luminescence, absorbance and droplet morphology [52]. For example, there has been a passive, size-based droplet separation device using deterministic lateral displacement (DLD), which takes advantage of the shrinkage of yeast-encapsulating droplets induced by the water efflux from these droplets to unoccupied droplets due to cell metabolism [52]. Hasan et al. [56] developed a droplet sorting approach based on the average fluorescence lifetimes of individual droplets. At a frequency of 40–50 Hz, they reliably sorted droplets containing either or both of the two dyes that were distinguishable by the average fluorescence lifetime [56].

(5) Another limitation of current droplet-based screening technologies is the requirement of a fluorogenic assay, which is not always available for many phenotypes. Gene expression profiling by mRNA sequencing is an alternative route for characterizing cell phenotypes. The emerging single-cell RNA sequencing droplet microfluidics (Drop-seq [96] and InDrop [97]) has the potential of being adapted for protein engineering in which the cell factories are usually bacterial and yeast cells that are much smaller than mammalian cells. In a pioneering study, Liu et al. [98] developed an isogenic colony sequencing (ICO-seq) system which integrated the expansion of yeast colonies in hydrogel microspheres with barcoded Drop-seq for high-throughput RNA sequencing, which can be used to characterize cellular phenotype for high-throughput screening applications.

(6) The capacity of droplet-based screening methods can be further boosted by leveraging the rapidly evolving technologies of artificial intelligence (AI), especially machine learning. Machine learning offers a route to enable automated monitoring of microfluidic systems by converting routinely collected sensor and image data into actionable information in real time [99]. It has been demonstrated very recently how machine learning-assisted image analysis can facilitate quality control over droplet generation [100] and efficiently code droplet populations [101]. With machine learning-supported image analysis, experimental conditions in microfluidic droplet assays can be encoded and decoded by colored beads [101]. Although it was not performed on droplets, a deep learning-assisted, image-activated cell sorting system was developed to sort in real time microalgal and blood cells based on intracellular protein localization and cell–cell interaction from large heterogeneous population [102].

Funding: This research received no external funding.

Conflicts of Interest: L.W. and J.E.S. are full-time employees of enEvolv, Inc., a synthetic biology company that designs and engineers microorganisms to produce chemicals, enzymes, and small molecules.

References

1. Chica, R.A. Protein engineering in the 21st century. *Protein Sci.* **2015**, *24*, 431. [CrossRef] [PubMed]
2. Yang, K.K.; Wu, Z.; Arnold, F.H. Machine-learning-guided directed evolution for protein engineering. *Nat. Methods* **2019**, *16*, 687–694. [CrossRef] [PubMed]
3. Chockalingam, K.; Blenner, M.; Banta, S. Design and application of stimulus-responsive peptide systems. *Protein Eng. Des. Sel.* **2007**, *20*, 155–161. [CrossRef] [PubMed]
4. Johannes, T.W.; Zhao, H. Directed evolution of enzymes and biosynthetic pathways. *Curr. Opin. Microbiol.* **2006**, *9*, 261–267. [CrossRef]
5. Schoemaker, H.E.; Mink, D.; Wubbolts, M.G. Dispelling the myths—Biocatalysis in industrial synthesis. *Science* **2003**, *299*, 1694–1697. [CrossRef]
6. Nielsen, J.; Keasling, J.D. Engineering cellular metabolism. *Cell* **2016**, *164*, 1185–1197. [CrossRef]
7. Wang, H.H.; Isaacs, F.J.; Carr, P.A.; Sun, Z.Z.; Xu, G.; Forest, C.R.; Church, G.M. Programming cells by multiplex genome engineering and accelerated evolution. *Nature* **2009**, *460*, 894. [CrossRef]
8. Jinek, M.; Chylinski, K.; Fonfara, I.; Hauer, M.; Doudna, J.A.; Charpentier, E. A programmable dual-RNA—Guided DNA endonuclease in adaptive bacterial immunity. *Science* **2012**, *337*, 816–821. [CrossRef]
9. Cong, L.; Ran, F.A.; Cox, D.; Lin, S.; Barretto, R.; Habib, N.; Hsu, P.D.; Wu, X.; Jiang, W.; Marraffini, L.A. Multiplex genome engineering using CRISPR/Cas systems. *Science* **2013**, *339*, 819–823. [CrossRef]
10. Vervoort, Y.; Linares, A.G.; Roncoroni, M.; Liu, C.; Steensels, J.; Verstrepen, K.J. High-throughput system-wide engineering and screening for microbial biotechnology. *Curr. Opin. Biotechnol.* **2017**, *46*, 120–125. [CrossRef]
11. Bjork, S.M.; Joensson, H.N. Microfluidics for cell factory and bioprocess development. *Curr. Opin. Biotechnol.* **2019**, *55*, 95–102. [CrossRef] [PubMed]
12. Stemmer, W.P. Rapid evolution of a protein in vitro by DNA shuffling. *Nature* **1994**, *370*, 389. [CrossRef] [PubMed]
13. Crameri, A.; Raillard, S.-A.; Bermudez, E.; Stemmer, W.P. DNA shuffling of a family of genes from diverse species accelerates directed evolution. *Nature* **1998**, *391*, 288. [CrossRef] [PubMed]
14. Moore, J.C.; Arnold, F.H. Directed evolution of a para-nitrobenzyl esterase for aqueous-organic solvents. *Nat. Biotechnol.* **1996**, *14*, 458. [CrossRef]
15. Romero, P.A.; Arnold, F.H. Exploring protein fitness landscapes by directed evolution. *Nat. Rev. Mol. Cell Biol.* **2009**, *10*, 866. [CrossRef]
16. Dougherty, M.J.; Arnold, F.H. Directed evolution: New parts and optimized function. *Curr. Opin. Biotechnol.* **2009**, *20*, 486–491. [CrossRef]
17. Tizei, A.G.; Csibra, E.; Torres, L.; Pinheiro, V.B. Selection platforms for directed evolution in synthetic biology. *Biochem. Soc. Trans.* **2016**, *44*, 1165. [CrossRef]
18. Bassalo, M.C.; Liu, R.; Gill, R.T. Directed evolution and synthetic biology applications to microbial systems. *Curr. Opin. Biotechnol.* **2016**, *39*, 126–133. [CrossRef]
19. Agresti, J.J.; Antipov, E.; Abate, A.R.; Ahn, K.; Rowat, A.C.; Baret, J.-C.; Marquez, M.; Klibanov, A.M.; Griffiths, A.D.; Weitz, D.A. Ultrahigh-throughput screening in drop-based microfluidics for directed evolution. *Proc. Natl. Acad. Sci. USA* **2010**, *107*, 4004–4009. [CrossRef]
20. Turner, N.J. Directed evolution drives the next generation of biocatalysts. *Nat. Chem. Biol.* **2009**, *5*, 567. [CrossRef]
21. Zinchenko, A.; Devenish, S.R.; Kintses, B.; Colin, P.-Y.; Fischlechner, M.; Hollfelder, F. One in a million: Flow cytometric sorting of single cell-lysate assays in monodisperse picolitre double emulsion droplets for directed evolution. *Anal. Chem.* **2014**, *86*, 2526–2533. [CrossRef] [PubMed]
22. Joensson, H.N.; Andersson-Svahn, H. Droplet microfluidics—A tool for protein engineering and analysis. *Lab Chip* **2011**, *11*, 4144–4147. [CrossRef] [PubMed]
23. Lipovšek, D.; Antipov, E.; Armstrong, K.A.; Olsen, M.J.; Klibanov, A.M.; Tidor, B.; Wittrup, K.D. Selection of horseradish peroxidase variants with enhanced enantioselectivity by yeast surface display. *Chem. Biol.* **2007**, *14*, 1176–1185. [CrossRef] [PubMed]

24. Aharoni, A.; Thieme, K.; Chiu, C.P.; Buchini, S.; Lairson, L.L.; Chen, H.; Strynadka, N.C.; Wakarchuk, W.W.; Withers, S.G. High-throughput screening methodology for the directed evolution of glycosyltransferases. *Nat. Methods* **2006**, *3*, 609. [CrossRef] [PubMed]

25. Dove, A. Screening for content—The evolution of high throughput. *Nat. Biotechnol.* **2003**, *21*, 859–864. [CrossRef] [PubMed]

26. Tawfik, D.S.; Griffiths, A.D. Man-made cell-like compartments for molecular evolution. *Nat. Biotechnol.* **1998**, *16*, 652. [CrossRef] [PubMed]

27. Thorsen, T.; Roberts, R.W.; Arnold, F.H.; Quake, S.R. Dynamic pattern formation in a vesicle-generating microfluidic device. *Phys. Rev. Lett.* **2001**, *86*, 4163. [CrossRef]

28. Brouzes, E. Droplet Microfluidics for Single-Cell Analysis. In *Single-Cell Analysis*; Springer: Berlin/Heidelberg, Germany, 2012; pp. 105–139.

29. Terekhov, S.S.; Smirnov, I.V.; Stepanova, A.V.; Bobik, T.V.; Mokrushina, Y.A.; Ponomarenko, N.A.; Belogurov, A.A.; Rubtsova, M.P.; Kartseva, O.V.; Gomzikova, M.O.; et al. Microfluidic droplet platform for ultrahigh-throughput single-cell screening of biodiversity. *Proc. Natl. Acad. Sci. USA* **2017**, *114*, 2550. [CrossRef]

30. Baret, J.-C.; Miller, O.J.; Taly, V.; Ryckelynck, M.; El-Harrak, A.; Frenz, L.; Rick, C.; Samuels, M.L.; Hutchison, J.B.; Agresti, J.J.; et al. Fluorescence-activated droplet sorting (FADS): Efficient microfluidic cell sorting based on enzymatic activity. *Lab Chip* **2009**, *9*, 1850–1858. [CrossRef]

31. Vallejo, D.; Nikoomanzar, A.; Paegel, B.M.; Chaput, J.C. Fluorescence-Activated Droplet Sorting for Single-Cell Directed Evolution. *ACS Synth. Biol.* **2019**, *8*, 1430–1440. [CrossRef]

32. Shang, L.; Cheng, Y.; Zhao, Y. Emerging droplet microfluidics. *Chem. Rev.* **2017**, *117*, 7964–8040. [CrossRef] [PubMed]

33. Joanicot, M.; Ajdari, A. Droplet control for microfluidics. *Science* **2005**, *309*, 887–888. [CrossRef] [PubMed]

34. Christopher, G.F.; Anna, S.L. Microfluidic methods for generating continuous droplet streams. *J. Phys. D* **2007**, *40*, R319. [CrossRef]

35. Gruner, P.; Riechers, B.; Chacòn Orellana, L.A.; Brosseau, Q.; Maes, F.; Beneyton, T.; Pekin, D.; Baret, J.-C. Stabilisers for water-in-fluorinated-oil dispersions: Key properties for microfluidic applications. *Curr. Opin. Colloid Interface Sci.* **2015**, *20*, 183–191. [CrossRef]

36. Mary, P.; Dauphinot, L.; Bois, N.; Potier, M.-C.; Studer, V.; Tabeling, P. Analysis of gene expression at the single-cell level using microdroplet-based microfluidic technology. *Biomicrofluidics* **2011**, *5*, 024109. [CrossRef] [PubMed]

37. Paegel, B.M.; Joyce, G.F. Microfluidic Compartmentalized Directed Evolution. *Chem. Biol.* **2010**, *17*, 717–724. [CrossRef]

38. Baret, J.-C. Surfactants in droplet-based microfluidics. *Lab Chip* **2012**, *12*, 422–433. [CrossRef]

39. Skhiri, Y.; Gruner, P.; Semin, B.; Brosseau, Q.; Pekin, D.; Mazutis, L.; Goust, V.; Kleinschmidt, F.; El Harrak, A.; Hutchison, J.B.; et al. Dynamics of molecular transport by surfactants in emulsions. *Soft Matter* **2012**, *8*, 10618–10627. [CrossRef]

40. Bibette, J.; Morse, D.C.; Witten, T.A.; Weitz, D.A. Stability criteria for emulsions. *Phys. Rev. Lett.* **1992**, *69*, 2439–2442. [CrossRef]

41. Abate, A.R.; Hung, T.; Mary, P.; Agresti, J.J.; Weitz, D.A. High-throughput injection with microfluidics using picoinjectors. *Proc. Natl. Acad. Sci. USA* **2010**, *107*, 19163. [CrossRef]

42. Ahn, K.; Agresti, J.; Chong, H.; Marquez, M.; Weitz, D.A. Electrocoalescence of drops synchronized by size-dependent flow in microfluidic channels. *Appl. Phys. Lett.* **2006**, *88*, 264105. [CrossRef]

43. Zagnoni, M.; Le Lain, G.; Cooper, J.M. Electrocoalescence Mechanisms of Microdroplets Using Localized Electric Fields in Microfluidic Channels. *Langmuir* **2010**, *26*, 14443–14449. [CrossRef] [PubMed]

44. Niu, X.; Gulati, S.; Edel, J.B.; de Mello, A.J. Pillar-induced droplet merging in microfluidic circuits. *Lab Chip* **2008**, *8*, 1837–1841. [CrossRef] [PubMed]

45. Akartuna, I.; Aubrecht, D.M.; Kodger, T.E.; Weitz, D.A. Chemically induced coalescence in droplet-based microfluidics. *Lab Chip* **2015**, *15*, 1140–1144. [CrossRef] [PubMed]

46. Duarte, J.M.; Barbier, I.; Schaerli, Y. Bacterial Microcolonies in Gel Beads for High-Throughput Screening of Libraries in Synthetic Biology. *ACS Synth. Biol.* **2017**, *6*, 1988–1995. [CrossRef] [PubMed]

47. Sjostrom, S.L.; Bai, Y.; Huang, M.; Liu, Z.; Nielsen, J.; Joensson, H.N.; Andersson Svahn, H. High-throughput screening for industrial enzyme production hosts by droplet microfluidics. *Lab Chip* **2014**, *14*, 806–813. [CrossRef]

48. Ma, F.; Chung, M.T.; Yao, Y.; Nidetz, R.; Lee, L.M.; Liu, A.P.; Feng, Y.; Kurabayashi, K.; Yang, G.-Y. Efficient molecular evolution to generate enantioselective enzymes using a dual-channel microfluidic droplet screening platform. *Nat. Commun.* **2018**, *9*, 1030. [CrossRef]

49. Fallah-Araghi, A.; Baret, J.-C.; Ryckelynck, M.; Griffiths, A.D. A completely in vitro ultrahigh-throughput droplet-based microfluidic screening system for protein engineering and directed evolution. *Lab Chip* **2012**, *12*, 882–891. [CrossRef]

50. Sciambi, A.; Abate, A.R. Adding reagent to droplets with controlled rupture of encapsulated double emulsions. *Biomicrofluidics* **2013**, *7*, 044112. [CrossRef]

51. Caen, O.; Schütz, S.; Jammalamadaka, M.S.S.; Vrignon, J.; Nizard, P.; Schneider, T.M.; Baret, J.-C.; Taly, V. High-throughput multiplexed fluorescence-activated droplet sorting. *Microsyst. Nanoeng.* **2018**, *4*, 33. [CrossRef]

52. Joensson, H.N.; Uhlén, M.; Svahn, H.A. Droplet size based separation by deterministic lateral displacement—Separating droplets by cell-induced shrinking. *Lab Chip* **2011**, *11*, 1305–1310. [CrossRef] [PubMed]

53. Frenzel, D.; Merten, C.A. Microfluidic train station: Highly robust and multiplexable sorting of droplets on electric rails. *Lab Chip* **2017**, *17*, 1024–1030. [CrossRef] [PubMed]

54. Pan, M.; Lyu, F.; Tang, S.K.Y. Fluorinated Pickering Emulsions with Nonadsorbing Interfaces for Droplet-based Enzymatic Assays. *Anal. Chem.* **2015**, *87*, 7938–7943. [CrossRef] [PubMed]

55. Pan, M.; Rosenfeld, L.; Kim, M.; Xu, M.; Lin, E.; Derda, R.; Tang, S.K.Y. Fluorinated Pickering Emulsions Impede Interfacial Transport and Form Rigid Interface for the Growth of Anchorage-Dependent Cells. *ACS Appl. Mater. Interfaces* **2014**, *6*, 21446–21453. [CrossRef]

56. Hasan, S.; Geissler, D.; Wink, K.; Hagen, A.; Heiland, J.J.; Belder, D. Fluorescence lifetime-activated droplet sorting in microfluidic chip systems. *Lab Chip* **2019**, *19*, 403–409. [CrossRef]

57. Eun, Y.-J.; Utada, A.S.; Copeland, M.F.; Takeuchi, S.; Weibel, D.B. Encapsulating Bacteria in Agarose Microparticles Using Microfluidics for High-Throughput Cell Analysis and Isolation. *ACS Chem. Biol.* **2011**, *6*, 260–266. [CrossRef]

58. Sandoz, P.A.; Chung, A.J.; Weaver, W.M.; Di Carlo, D. Sugar Additives Improve Signal Fidelity for Implementing Two-Phase Resorufin-Based Enzyme Immunoassays. *Langmuir* **2014**, *30*, 6637–6643. [CrossRef]

59. Courtois, F.; Olguin, L.F.; Whyte, G.; Theberge, A.B.; Huck, W.T.S.; Hollfelder, F.; Abell, C. Controlling the Retention of Small Molecules in Emulsion Microdroplets for Use in Cell-Based Assays. *Anal. Chem.* **2009**, *81*, 3008–3016. [CrossRef]

60. Brouzes, E.; Medkova, M.; Savenelli, N.; Marran, D.; Twardowski, M.; Hutchison, J.B.; Rothberg, J.M.; Link, D.R.; Perrimon, N.; Samuels, M.L. Droplet microfluidic technology for single-cell high-throughput screening. *Proc. Natl. Acad. Sci. USA* **2009**, *106*, 14195–14200. [CrossRef]

61. Nightingale, A.M.; Phillips, T.W.; Bannock, J.H.; de Mello, J.C. Controlled multistep synthesis in a three-phase droplet reactor. *Nat. Commun.* **2014**, *5*, 3777. [CrossRef]

62. Eow, J.S.; Ghadiri, M. Drop-drop coalescence in an electric field: The effects of applied electric field and electrode geometry. *Colloids Surf. A* **2003**, *219*, 253–279. [CrossRef]

63. Song, H.; Li, H.-W.; Munson, M.S.; Van Ha, T.G.; Ismagilov, R.F. On-Chip Titration of an Anticoagulant Argatroban and Determination of the Clotting Time within Whole Blood or Plasma Using a Plug-Based Microfluidic System. *Anal. Chem.* **2006**, *78*, 4839–4849. [CrossRef] [PubMed]

64. Um, E.; Lee, D.-S.; Pyo, H.-B.; Park, J.-K. Continuous generation of hydrogel beads and encapsulation of biological materials using a microfluidic droplet-merging channel. *Microfluid. Nanofluid.* **2008**, *5*, 541–549. [CrossRef]

65. Bernath, K.; Hai, M.; Mastrobattista, E.; Griffiths, A.D.; Magdassi, S.; Tawfik, D.S. In vitro compartmentalization by double emulsions: Sorting and gene enrichment by fluorescence activated cell sorting. *Anal. Biochem.* **2004**, *325*, 151–157. [CrossRef] [PubMed]

66. Mastrobattista, E.; Taly, V.; Chanudet, E.; Treacy, P.; Kelly, B.T.; Griffiths, A.D. High-Throughput Screening of Enzyme Libraries: In Vitro Evolution of a β-Galactosidase by Fluorescence-Activated Sorting of Double Emulsions. *Chem. Biol.* **2005**, *12*, 1291–1300. [CrossRef] [PubMed]

67. Miller, O.J.; Bernath, K.; Agresti, J.J.; Amitai, G.; Kelly, B.T.; Mastrobattista, E.; Taly, V.; Magdassi, S.; Tawfik, D.S.; Griffiths, A.D. Directed evolution by in vitro compartmentalization. *Nat. Methods* **2006**, *3*, 561–570. [CrossRef] [PubMed]

68. Aharoni, A.; Amitai, G.; Bernath, K.; Magdassi, S.; Tawfik, D.S. High-Throughput Screening of Enzyme Libraries: Thiolactonases Evolved by Fluorescence-Activated Sorting of Single Cells in Emulsion Compartments. *Chem. Biol.* **2005**, *12*, 1281–1289. [CrossRef]

69. Amsden, B. Solute Diffusion within Hydrogels. Mechanisms and Models. *Macromolecules* **1998**, *31*, 8382–8395. [CrossRef]

70. Wechsler, M.E.; Stephenson, R.E.; Murphy, A.C.; Oldenkamp, H.F.; Singh, A.; Peppas, N.A. Engineered microscale hydrogels for drug delivery, cell therapy, and sequencing. *Biomed. Microdevices* **2019**, *21*, 31. [CrossRef]

71. Cha, C.; Oh, J.; Kim, K.; Qiu, Y.; Joh, M.; Shin, S.R.; Wang, X.; Camci-Unal, G.; Wan, K.-T.; Liao, R. Microfluidics-assisted fabrication of gelatin-silica core-shell microgels for injectable tissue constructs. *Biomacromolecules* **2014**, *15*, 283–290. [CrossRef]

72. Ahn, K.; Kerbage, C.; Hunt, T.P.; Westervelt, R.M.; Link, D.R.; Weitz, D.A. Dielectrophoretic manipulation of drops for high-speed microfluidic sorting devices. *Appl. Phys. Lett.* **2006**, *88*, 024104. [CrossRef]

73. Fidalgo, L.M.; Whyte, G.; Bratton, D.; Kaminski, C.F.; Abell, C.; Huck, W.T. From microdroplets to microfluidics: Selective emulsion separation in microfluidic devices. *Angew. Chem. Int. Ed.* **2008**, *47*, 2042–2045. [CrossRef] [PubMed]

74. Chen, Y.; Wijaya Gani, A.; Tang, S.K. Characterization of sensitivity and specificity in leaky droplet-based assays. *Lab Chip* **2012**, *12*, 5093–5103. [PubMed]

75. Woronoff, G.; El Harrak, A.; Mayot, E.; Schicke, O.; Miller, O.J.; Soumillion, P.; Griffiths, A.D.; Ryckelynck, M. New Generation of Amino Coumarin Methyl Sulfonate-Based Fluorogenic Substrates for Amidase Assays in Droplet-Based Microfluidic Applications. *Anal. Chem.* **2011**, *83*, 2852–2857. [PubMed]

76. Ramsden, W. Separation of solids in the surface-layers of solutions and 'suspensions' (observations on surface-membranes, bubbles, emulsions, and mechanical coagulation)—Preliminary account. *Proc. R. Soc. Lond.* **1904**, *72*, 156–164.

77. Pickering, S.U. Cxcvi—Emulsions. *J. Chem. Soc. Trans.* **1907**, *91*, 2001–2021.

78. Dinsmore, A.; Hsu, M.F.; Nikolaides, M.; Marquez, M.; Bausch, A.; Weitz, D. Colloidosomes: Selectively permeable capsules composed of colloidal particles. *Science* **2002**, *298*, 1006–1009. [CrossRef]

79. Subramaniam, A.B.; Abkarian, M.; Stone, H.A. Controlled assembly of jammed colloidal shells on fluid droplets. *Nat. Mater.* **2005**, *4*, 553.

80. Crossley, S.; Faria, J.; Shen, M.; Resasco, D.E. Solid nanoparticles that catalyze biofuel upgrade reactions at the water/oil interface. *Science* **2010**, *327*, 68–72. [CrossRef]

81. Vertegel, A.A.; Siegel, R.W.; Dordick, J.S. Silica Nanoparticle Size Influences the Structure and Enzymatic Activity of Adsorbed Lysozyme. *Langmuir* **2004**, *20*, 6800–6807.

82. Czeslik, C.; Winter, R. Effect of temperature on the conformation of lysozyme adsorbed to silica particles. *Phys. Chem. Chem. Phys.* **2001**, *3*, 235–239. [CrossRef]

83. Preari, M.; Spinde, K.; Lazic, J.; Brunner, E.; Demadis, K.D. Bioinspired Insights into Silicic Acid Stabilization Mechanisms: The Dominant Role of Polyethylene Glycol-Induced Hydrogen Bonding. *J. Am. Chem. Soc.* **2014**, *136*, 4236–4244. [CrossRef] [PubMed]

84. Pan, M.; Lyu, F.; Tang, S.K.Y. Methods to coalesce fluorinated Pickering emulsions. *Anal. Methods* **2017**, *9*, 4622–4629. [CrossRef]

85. Siltanen, C.A.; Cole, R.H.; Poust, S.; Chao, L.; Tyerman, J.; Kaufmann-Malaga, B.; Ubersax, J.; Gartner, Z.J.; Abate, A.R. An oil-free picodrop bioassay platform for synthetic biology. *Sci. Rep.* **2018**, *8*, 7913. [CrossRef] [PubMed]

86. Cole, R.H.; Tang, S.-Y.; Siltanen, C.A.; Shahi, P.; Zhang, J.Q.; Poust, S.; Gartner, Z.J.; Abate, A.R. Printed droplet microfluidics for on demand dispensing of picoliter droplets and cells. *Proc. Natl. Acad. Sci. USA* **2017**, *114*, 8728–8733. [CrossRef]

87. Sinha, R.; Shukla, P. Current Trends in Protein Engineering: Updates and Progress. *Curr. Protein Pept. Sci.* **2019**, *20*, 398–407. [CrossRef]

88. Available online: https://www.zionmarketresearch.com/report/protein-engineering-market (accessed on 14 February 2019).

89. Diefenbach, X.W.; Farasat, I.; Guetschow, E.D.; Welch, C.J.; Kennedy, R.T.; Sun, S.; Moore, J.C. Enabling biocatalysis by high-throughput protein engineering using droplet microfluidics coupled to mass spectrometry. *ACS Omega* **2018**, *3*, 1498–1508. [CrossRef]

90. El Debs, B.; Utharala, R.; Balyasnikova, I.V.; Griffiths, A.D.; Merten, C.A. Functional single-cell hybridoma screening using droplet-based microfluidics. *Proc. Natl. Acad. Sci. USA* **2012**, *109*, 11570–11575. [CrossRef]

91. Liu, X.; Painter, R.; Enesa, K.; Holmes, D.; Whyte, G.; Garlisi, C.; Monsma, F.; Rehak, M.; Craig, F.; Smith, C.A. High-throughput screening of antibiotic-resistant bacteria in picodroplets. *Lab Chip* **2016**, *16*, 1636–1643. [CrossRef]

92. Edd, J.F.; Di Carlo, D.; Humphry, K.J.; Köster, S.; Irimia, D.; Weitz, D.A.; Toner, M. Controlled encapsulation of single-cells into monodisperse picolitre drops. *Lab Chip* **2008**, *8*, 1262–1264. [CrossRef]

93. Lagus, T.P.; Edd, J.F. High-throughput co-encapsulation of self-ordered cell trains: Cell pair interactions in microdroplets. *RSC Adv.* **2013**, *3*, 20512–20522. [CrossRef]

94. Mutlu, B.R.; Edd, J.F.; Toner, M. Oscillatory inertial focusing in infinite microchannels. *Proc. Natl. Acad. Sci. USA* **2018**, *115*, 7682–7687. [CrossRef] [PubMed]

95. Cruz, J.; Graells, T.; Walldén, M.; Hjort, K. Inertial focusing with sub-micron resolution for separation of bacteria. *Lab Chip* **2019**, *19*, 1257–1266. [CrossRef] [PubMed]

96. Macosko, E.Z.; Basu, A.; Satija, R.; Nemesh, J.; Shekhar, K.; Goldman, M.; Tirosh, I.; Bialas, A.R.; Kamitaki, N.; Martersteck, E.M.; et al. Highly Parallel Genome-wide Expression Profiling of Individual Cells Using Nanoliter Droplets. *Cell* **2015**, *161*, 1202–1214. [CrossRef]

97. Klein, A.M.; Mazutis, L.; Akartuna, I.; Tallapragada, N.; Veres, A.; Li, V.; Peshkin, L.; Weitz, D.A.; Kirschner, M.W. Droplet Barcoding for Single-Cell Transcriptomics Applied to Embryonic Stem Cells. *Cell* **2015**, *161*, 1187–1201. [CrossRef]

98. Liu, L.; Dalal, C.K.; Heineike, B.M.; Abate, A.R. High throughput gene expression profiling of yeast colonies with microgel-culture Drop-seq. *Lab Chip* **2019**, *19*, 1838–1849. [CrossRef]

99. Riordon, J.; Sovilj, D.; Sanner, S.; Sinton, D.; Young, E.W.K. Deep Learning with Microfluidics for Biotechnology. *Trends Biotechnol.* **2019**, *37*, 310–324. [CrossRef]

100. Chu, A.; Nguyen, D.; Talathi, S.S.; Wilson, A.C.; Ye, C.; Smith, W.L.; Kaplan, A.D.; Duoss, E.B.; Stolaroff, J.K.; Giera, B. Automated detection and sorting of microencapsulation via machine learning. *Lab Chip* **2019**, *19*, 1808–1817. [CrossRef]

101. Svensson, C.-M.; Shvydkiv, O.; Dietrich, S.; Mahler, L.; Weber, T.; Choudhary, M.; Tovar, M.; Figge, M.T.; Roth, M. Coding of Experimental Conditions in Microfluidic Droplet Assays Using Colored Beads and Machine Learning Supported Image Analysis. *Small* **2019**, *15*, 1802384. [CrossRef]

102. Nitta, N.; Sugimura, T.; Isozaki, A.; Mikami, H.; Hiraki, K.; Sakuma, S.; Iino, T.; Arai, F.; Endo, T.; Fujiwaki, Y.; et al. Intelligent Image-Activated Cell Sorting. *Cell* **2018**, *175*, 266–276. [CrossRef]

MDPI

St. Alban-Anlage 66

4052 Basel

Switzerland

Tel. +41 61 683 77 34

Fax +41 61 302 89 18

www.mdpi.com

Micromachines Editorial Office

E-mail: micromachines@mdpi.com

www.mdpi.com/journal/micromachines